AF212571

Mobile Tactical Data Networks

S|D|G
System Design Group

Mobile Tactical Data Networks

Ian Brace, PhD

System Design Group Pty Ltd
Australia 2014

System Design Group Pty Ltd, Mornington, VIC Australia
©2014 System Design Group Pty Ltd. All rights reserved.

National Library of Australia Cataloguing-in-Publication entry

---

Brace, Ian, author.

Mobile tactical data networks / Ian Brace.

ISBN: 9780992529901 (hardback)
Includes bibliographical references and index.
Tactics–Data processing.
Military art and science–Data processing.
Electronics in military engineering.
Information warfare.
355.07

Cover design by Shelly Media.

About the Author

Dr Ian Brace is an electrical engineer with twenty years of experience in the Defence domain. He served in the Australian Defence Force for many years, specialising in communication systems and electronic warfare, and received his engineering degree from the University of New South Wales. He completed the Master of Science degree at the Royal Military College of Science in the United Kingdom, specialising in military electronic systems engineering for radar and communication systems, and he has consulted in model-based systems engineering for the last decade. Dr Brace was awarded the PhD in 2006 from the University of Melbourne where his research explored the application of differential geometry to a range of signal processing problems. He lives in Mornington, Victoria.

Contents

Preface

This book is based on a five day training course on mobile tactical data networks, and it is designed to introduce engineers to the particular challenges of establishing and maintaining tactical networks on the battlefield. In the past it was military research that invented clever communication techniques which were then picked up by the commercial players and converted into even better civilian networks. The internet itself is a military invention, along with spread spectrum technology which is the core technology in many telephone networks and wireless systems. However, in recent times it seems that it is the military that is lagging behind the civilian players, with video applications streaming to my iPhone while I am driving in a car, but tactical data networks strolling along with one tenth of the capacity.

As the civilian infrastructure develops it is easy to forget the unique challenges faced on the battlefield. There is no infrastructure, except what the military force brings with it; there are no cell towers; there is no fibre to the premises. The end users are typically hiding behind the hill, rather than sitting up on top of it, and they can be a few kilometres apart or hundreds of kilometres apart. Networks are ad-hoc and mobile and there is an adversary deliberately trying to jam the signals. Although the technology and the services are similar to their civilian counterpart, establishing and maintaining these networks is an order of magnitude more difficult.

Despite these problems the role of data on the battlefield is now well established. This book is designed to highlight why military data communications are so difficult and what has been done to try and mitigate against these difficulties. To that end we will draw our examples from real systems, examining some of the benchmark systems that have been deployed or are coming into service. In many ways the United States has set the standards for military communications, so they provide good examples of typical technologies. However, the book also discusses the work of other nations and organisations.

Chapter one introduces tactical communications by setting it in a battlefield context. The communication system is there to support commanders in the planning and execution of operations and it must accommodate the commanders battlefield requirements, not constrain his tactical options. We look at the evolution of the communication system as military thinking has evolved from C3 to C4ISR. Chapter two reviews the basic digital communication theory required for wireless data communications and introduces some of the difficulties in implementing these features in the tactical domain. Each of the techniques discussed in this chapter are subsequently found in the example radios and data links examined in the remainder of the book. By understanding these techniques and the role they perform one can better understand what the various features in each radio are trying to do.

In chapter three we examine legacy tactical data architectures, such as Link 16, SINCGARS and EPLRS, by explaining the evolution of these non-IP data networks that have been successfully deployed for many years. These systems are still in widespread use today and are likely to remain so for some time to come since the new generation of radios are generally backwards compatible with these older systems. The transition of these systems from non-IP isolated networks to being part of the Tactical Internet architecture is also reviewed, as part of the evolution of data networks. Chapter four then introduces the current approach to tactical data networks as major military forces are transitioning to an all-IP infrastructure. This chapter examines the United States JTRS program and the new waveforms that have been defined through it, as well as the trend towards a Software Communications Architecture. We also examine some of the current commercial offerings that have found service in Iraq and Afghanistan as real-world examples of the current approach to data.

For modern military forces the brigade-level formation has become the nexus between the lower and upper tactical domain and is therefore a logical place from which to consider the unique communications needs of brigade and below. In a high mobility future battlefield it seems likely that dispersion will become a key tactic to defeat the effects of accurate stand-off weapons and permanent surveillance systems. Therefore, chapter five considers beyond-line-of-sight communications, with a strong emphasis on satellite communications and in particular using satellite services to support data to mobile operations. Chapter six complements this perspective by considering "data below brigade" by returning to combat net radio as a data bearer and considering the role of battle management systems for modern forces. Data over high frequency radio is also briefly addressed.

As the complexity of modern battlefield communications increases, fitting all of these systems into a military vehicle becomes more challenging. Chapter seven addresses some of the common installation issues that face modern forces, such as radiation hazards to personnel, power requirements, space and security constraints and electromagnetic interference. The emerging Vetronics open architecture for vehicle electronics is also discussed. Finally, the military electromagnetic environment is hostile so chapter eight considers the EW threat and describes the basic countermeasures available, and how modern waveforms using spread spectrum techniques can improve network resilience.

Ian M. Brace

1 C4ISR in Support of Mobile Military Operations

1.1 Introduction

COMMUNICATIONS technologies, particularly data systems, are changing the way that armies fight on the battlefield. The changes that started in the early nineties have accelerated over the last two decades to produce tactical data networks that can support battle management services, imagery, video, location data, voice command and surveillance streaming. Much of this progress has evolved from similar progress in commercial communication systems where mobile networks are built with large, stable, fixed base stations that implement static network coverage in a single fixed geographic area. Such stations can be built with well designed antennae and air-conditioned cabinets in order to achieve stable, continuous high data rates over a fixed frequency band. The end users are typically close to the base station with line-of-sight access from their phone to the station. Base stations provide height, which improves the likelihood of a successful telephone call. Whilst some of the advances from commercial systems have been imported into military networks, it has also been necessary to develop specific novel solutions because of the distinctly different operating conditions that deployed forces face.

Tactical military networks are often ad-hoc and are usually established in harsh climatic conditions. They are typically peer-to-peer networks and the distance between members of the network can be several kilometres or even hundreds of kilometres. For tactical networks on the move, base stations are impractical and the members of the network often need to avoid high, exposed locations. Hence, the terrain can often shield direct radio line-of-sight between members of the network which will degrade the link performance. The electromagnetic environment can also be hostile, with an adversary trying to disrupt the force communications.

Military vehicles travel at speed over rough terrain whilst under fire from enemy forces. The systems must be physically robust against heat, dust, rain and shock. The vehicle's primary mission is not usually to carry the communications suite, so space is a premium. Multiple systems must be installed in a small space and must operate effectively without interfering with other electronic systems. Since a vehicle is essentially a metal box on wheels (or tracks), electronic earthing is essential, and the confined crew cabin can also create a radio frequency radiation hazard.

Given the challenges facing tactical communications systems it is essential that we understand the role of these systems in supporting modern forces to fight. By clearly understanding what the war fighter needs from the communications system we can focus our efforts on delivering the right service to the right person at the right time.

1.1.1 [R]Evolution in Military Affairs

In the early 1990s, as the internet was flourishing and the first mobile phones appeared, there was a sense of expectation that the dawning information age would fundamentally change the way that warfare would be executed in the future. The term *Revolution in Military Affairs* (RMA) was used to capture this idea and there was a resurgence of popularity in books like Alvin Toffler's *Future Shock* [5], which defines the development of the nation state through agrarian, industrial and post-industrial stages. He argued that this post-industrial stage is an information age and coined the term *information overload.* It was suggested that the RMA would fundamentally alter tactical warfare around *Information Operations* [1], and the side that won the information battle would win the encounter.

In practice, *Information Operations* tend to be implemented at the strategic or operational level, with tactical troops providing the force aspect of the plan. In the late 1990s there was a shift in focus as military forces looked for strategies that could implement the revolution at the tactical level, and the *Network Centric Warfare* (NCW) concept gained currency. NCW emphasises the use of data networks to speed up the actions and response of forces on the battlefield. The possibility of real-time sensor to shooter links could bring weapons to bear on an adversary so quickly that he is unable to react or respond. Whereas the communication system had previously been seen as a tool for command and control (referred to as *C3*), NCW aspired to integrate intelligence and surveillance assets, as well as target acquisition systems, into a single tactical data network that could act as a force multiplier (referred to as *C4ISR*).

Over the last decade there have been significant developments in establishing a ubiquitous tactical data network. However, with the benefit of hindsight it has been suggested that there has been, perhaps, an evolution in military affairs rather than a revolution [3]. For example, traditional concealment is less effective against modern integrated surveillance systems, which has led to a greater reliance on mobility to avoid detection. Similarly, traditional cover has become less effective due to advances in the lethality and stand-off capabilities of modern weapons, which has led to an emphasis on protection and mobility. If future fighting forces must be highly mobile with relatively superior situational awareness, access to surveillance feeds and integrated targeting systems, then tactical communication systems must provide ad-hoc, high bandwidth, vehicle mounted data networks that operate on the move.

1.1.2 Military Acquisition Programs

Legacy military communication systems have a significant impact on the development of new networks. The armed forces of each modern country have invested heavily in communication projects over the last twenty years and some of these programs have run for up to ten years. Unfortunately, the military has had a history of creating turnkey solutions for communications which then take too many years to develop and deploy. The result has been the delivery of high quality equipment that addresses the communication needs that were identified five years before. Of course, modern systems are evolving quickly and the current needs are often quite different, but after a large outlay of funds there is little opportunity to re-design the system.[1]

Because there are many widely deployed legacy assets, engineers have sought ways to integrate legacy equipment into modern networks. However, to understand the limitations of legacy equipment it is helpful to review the path that military communications has travelled during the last twenty years.

1.2 Analogue Voice Command and Control

In the mid-1980s military forces still concerned themselves with C3 - Command, Control and Communications. Tactical command and control of most operations was exercised through insecure radio voice nets, organised in a hierarchical structure to mirror the chain of command. Tactical operations were planned at Division Headquarters (HQ) and implemented by subordinate

[1] See reference [3] for a discussion about the current UK and Canadian situation.

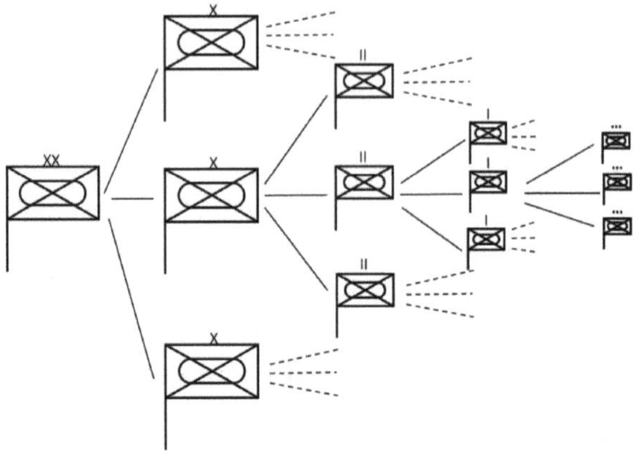

Figure 1.2.1: Divisional Command Structure

brigades, as shown in figure 1.2.1. Control of scarce intelligence or surveillance assets would be held at Division HQ and staff compile the various inputs into a coherent operational picture. The operational situation would then be delivered to the subordinate formation commanders and their staff. Delivering "high bandwidth" data would be done through regular face-to-face meetings or briefings, in which the commanders would each travel to Division HQ, or sometimes using a dispatch rider to courier maps and reports forward to the subordinate formations.

Communications at that time was divided into two distinct subsystems: the trunk communication subsystem and the combat net radio (CNR) subsystem. The trunk communication subsystem provided higher capacity UHF bearers between Division HQ and the three subordinate Brigade HQ (which comprised Brigade HQ Main and Brigade HQ Rear, where the support and logistic elements were stationed), which supported a field telephone system and a hard copy message system. Within a brigade, all communications from the Brigade HQ to each battalion was implemented using the combat net radio subsystem. These circuits comprised a single channel radio with each channel dedicated to a particular purpose. A brigade would typically run two voice circuits and one hard copy message circuit, with the two voice circuits designated as the brigade command net and the brigade administration net. The command net was the primary operational communications channel because it was the channel used by the commander to direct his force once battle was under way. Even today, providing a voice command channel continues to be a primary communications function of the CNR subsystem. Integrating voice with data, but ensuring voice always has primacy, is an ingoing operational requirement.

1.3 The Digitised Battlefield

1.3.1 Introduction

Digitisation came to the battlefield in the 1980s as analogue voice systems gave way to radios that were designed for digital waveforms. The driver towards digitised voice was security, because voice encoders could produce a digital data stream to which sophisticated encryption algorithms could then be applied. Digital radios, such as SINCGARS, were equipped with frequency shift keying and phase shift keying modulation schemes and initially they simply replaced the existing analogue voice radios.

In the early 1990s there was a growing need to improve situational awareness across the battlefield and digital data messages were the obvious candidate. Air and maritime forces had successfully employed Link 16 messaging to provide an enhanced operational picture to all elements on the Link 16 network and a similar concept was desirable in the land domain. However, digital hierarchical radio nets that mirror the chain of command were not effective in transmitting situational awareness data.

1.3.2 Situational Awareness

Situational awareness is the extent to which a commander knows the disposition of adversary, friendly, neutral and non-combatant entities across the battlefield. Situational awareness needs to be *real time* or *near-real time* in order to support the commander in achieving decision superiority over the adversary. Hierarchical network structures cannot provide the required bandwidth to support *near-real time* situational data transfer.

Consider the example of a mechanised brigade in which the fundamental unit is a mechanised platoon of four vehicles [4]. *Near-real time* situational awareness requires that the location of every vehicle is known at all times. The accuracy of the position location should be within at least ten metres for battlefield identification. A location report is likely to require at least 100 bytes of data, which would become 200 bytes after redundant bits are added to support error detection and correction.

If the platoon advances across country at 40 km/hr (11ms^{-1}) then each vehicle must transmit a location report at least every second to maintain location data accurate to within ten metres.The platoon net would therefore require a capacity of $4 \times 200 \times 8 = 6.4$ kbps; the company net would require a capacity of 25.6 kbps; the battalion net would require 102.4 kbps; and the brigade net would require 500 kbps. However, combat net radio was struggling to provide more than a few kbps, which were quickly overwhelmed in a hierarchical net structure. Clearly the solution was to provide a tactical data network with a meshed topology that could provide multiple routes through the network, but in the harsh tactical wireless environment there were few solutions available.

1.3.3 Tactical Data

In the mid-nineties the US military launched the SINCGARS System Improvement Program (SIP) which produced a new SINCGARS radio with a built-in point-to-point data capability nominally providing a 600 - 4800 bps data rate. The Internet Network Controller (INC) was added, which provided the first tactical packet data capability over combat net radio. The INC used a store-and-forward capability to push out packets while there was no voice activity on the network [2]. In 1999 the Advanced System Improvement Program (ASIP) released a further upgrade, which used a two-channel TDMA-type arrangement to transmit voice and data simultaneously.

Following the Gulf War in 1991, the success of the Position Location Reporting System (PLRS) launched an improvement program to create the Enhanced PLRS (EPLRS) which could provide the radio backbone for a formation level tactical data network. At the same time, other modern armies were also attempting various digitisation strategies. The Australian Army introduced the Battlefield Command Support System (BCSS), which provided a basic battlefield management application, but used the existing analogue combat net radio as the bearer. The traditional brigade administration net was subsumed into the new data net and additional computers, modems and GPS equipment were fitted to the armoured vehicle fleet. These solutions were slow, but provided a starting place for digital experimentation. With overhead due to encryption and error correction codes the genuine data throughput could be as low as tens of bits per second.

The first decade of the twenty first century has seen steady progress in the development of tactical digital systems. The US committed to developing a family of software defined radios that could be employed across all arms of the military. The backbone of the US effort has been the Joint Tactical Radio System (JTRS); an ambitious program to develop a family of software defined radios that would integrate all of the communications needs of the disparate arms of the US military. The JTRS has developed several non-proprietary waveforms, as well as defining a standard architecture for the radio family. Unfortunately, as with many large defence programs, the JTRS slipped behind schedule and over budget. After ten years and several billion dollars the program was eventually stopped when trials of the first Ground Mounted Radio were disappointing. The trial was conducted in 2011 but was a failure and finally the program was cancelled in 2012. Nevertheless, several commercial companies took advantage of the program's delay to produce their own narrow and wide band tactical radios to meet the existing military needs. The Raytheon EPLRS has been the de facto standard wideband line of sight tactical data network for many years, but in the mid-2000s it was modified to take greater advantage of its IP networking capability. Harris has also deployed a family of digital radios; the AN/PRC 117G wideband radio was widely used in Afghanistan and the AN/PRC 152 is a popular narrowband tactical radio. These systems provide genuine networks and can pass data between similar devices and also across network boundaries, which breaks the traditional nexus between hierarchical communication networks and hierarchical command structures. These Internet Protocol networks can carry the standard range of data services, including imagery and some video. In keeping with this development the traditional C3 gave way to the more contemporary C4ISR (Command, Control, Communications, Computing, Intelligence, Surveillance and Reconnaissance).

1.4 Future Warfare and Systems

Modern armies, such as the US, UK and Canada, appear to see future warfare focussed at brigade level, with protected mobile formations executing more autonomously than in the past [3]. The requirement for increased mobility and protection is in response to improved surveillance and target acquisition systems. Greater autonomy implies an increased need for current surveillance and intelligence data at brigade level, including greater use of autonomous drones and other sensor systems. With this in mind, modern tactical communication systems can be characterised by four key attributes:

- **Capacity.** Can the system deliver the necessary information to the re-

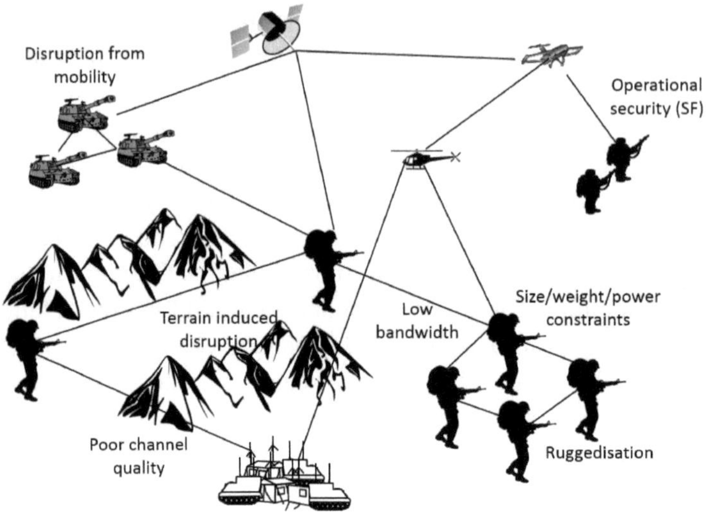

Figure 1.3.1: Tactical Data Constraints

quired location within the required time frame?

- **Reach.** Can the system push enhanced capacity as far forward as required - to battalion or company level?
- **Mobility.** Can the system deliver sufficient capacity whilst tactical elements are on the move?
- **Availability.** Can the system reliably support ad-hoc networks with redundant paths?

Table 1.1 is an estimate of the future capacity of military communications for three modern armies, based on publicly available data on their existing acquisition programs [3]. Of the three countries, only the United States appears likely to achieve the sort of capacity that will be required in order to conduct mobile, autonomous brigade operations as envisaged by the doctrine of each country. However, this has come at an enormous cost in research and development. Current efforts have shifted towards defining the tactical waveforms to be used in military radios, hence allowing a range of companies to produce compliant platforms.

	JTF HQ	Bde HQ (Static)	Bde HQ (Mobile)	Bn HQ	Coy HQ
US	16 Mbps	160 Mbps	60 Mbps	70 Mbps	1 Mbps
UK	13 Mbps	140 Mbps	300 kbps	400 kbps	400 kbps
Canada	9 Mbps	60 Mbps	600 kbps	600 kbps	600 kbps

Table 1.1: Projected tactical data rates for the US, UK and Canadian military

1.5 Conclusion

The acronym C4ISR has evolved from the original C3 to reflect the way that the current digital tactical communication systems have evolved from analogue voice radio nets. The future goal in this evolution is to integrate voice, data, imagery and video into a single "tactical cloud" that delivers the right information to the right person in the right place within the right time frame. The challenge is to do that on ad-hoc networks without fixed infrastructure in a mobile environment with a hostile electromagnetic opponent.

The remainder of this book will discuss some of the key challenges in achieving that goal, as well as outlining some of the solutions that are currently available. In order to do this we need to cover some of the basics around digital communications at the lower layers in order to understand why the evolutionary road always seems so steep. Then we will consider a range of options for delivering digital data to the battlefield, including terrestrial and satellite solutions, followed by the key issues involved in installing secure vehicle based systems.

References

[1] ADF. *Australian Defence Doctrine Publication 3.13 Information Operations*. Defence Publishing Service Canberra 2600, 2006.

[2] US Army. *FM 6-02.53 Tactical Radio Operations*. Department of Defense, 2009.

[3] L Hogan. G. Petch, P Straughair. Occasional paper: The impact of tactical communications on future military effectiveness. Technical report, Raytheon Australia Pty Ltd, 2012.

[4] M. Ryan and M. Frater. *Battlefield Communications Systems.* Argos Press, 2007.

[5] Alvin Toffler. *Future Shock.* Random House, 1970.

2 Digital Techniques in the Tactical Domain

2.1 The Tactical Network Model

2.1.1 A Standard Network Model

SINCE the second world war, commanders have fought battles using radio voice subnets to communicate with their troops, and relied on vehicle based radio systems to link them with higher headquarters. These higher links were typically radio relay, multi-channel, circuit switched systems, providing battlefield telephone systems and messaging. Whilst updated versions of this functionality are still deployed, this type of communication system is a legacy system when compared to the current and future battlefield requirements. The last decade has established data communications as the primary technical approach to battlefield management, although voice communications retains primacy for battle command. The rise in commercial Internet Protocol (IP) systems has crossed over into the tactical domain, such that IP networking has become a key requirement for tactical communication systems. Since the mid-2000s all major defence communications vendors have been striving to deploy IP enabled networking products to military forces.

Any discussion of data networking is usually based around a protocol stack model that segments the various requirements needed to complete a successful end-to-end transmission. The Open Systems Interconnection (OSI) model is an industry standard model based around seven layers of functionality. Each layer in the model on the host machine communicates with its corresponding layer on a destination machine, as illustrated in Figure 2.1.1. Data flows up and down the protocol stack using well defined interfaces between the layers.

The role of each layer in the OSI model can be summarised as:

1. Application Layer. A software process that performs its intended application, such as an email client or web browser.

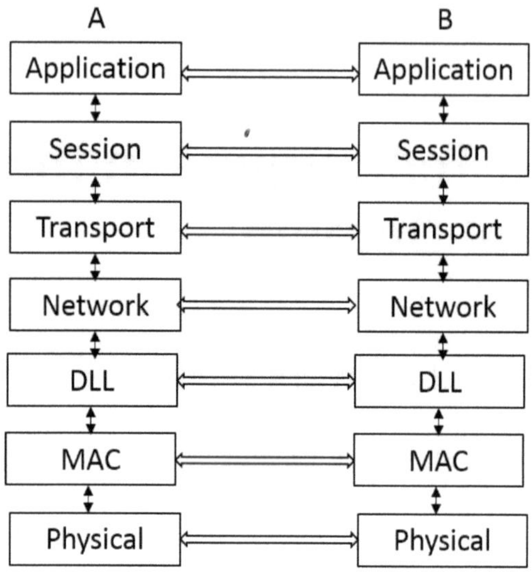

Figure 2.1.1: The Layers of the OSI Model

2. Session Layer. Typically used for authentication, or access rights, such as the management of Secure Sockets Layer (SSL) encryption.
3. Transport Layer. An end-to-end peer process that:
   a) breaks a message into packets at the sender and accounts for them all arriving at the receiver,
   b) manages the flow control requirements on a congested end-to-end path,
   c) manages session multiplexing if there are several low rate sessions between the same node pair.
4. Network Layer. A peer-to-peer protocol that manages packet flow control and routing across the network. To achieve this the network layer generates its own control packets, such as Link State Updates and route discovery packets.
5. Data Link Layer (DLL). A peer process that ensures the reliability of the physical link from the host to the next peer in the network. Different DLL protocols can include different error detection or correction approaches. There is debate about how much error control should reside in the DLL layer, as opposed to relying on the transport layer.

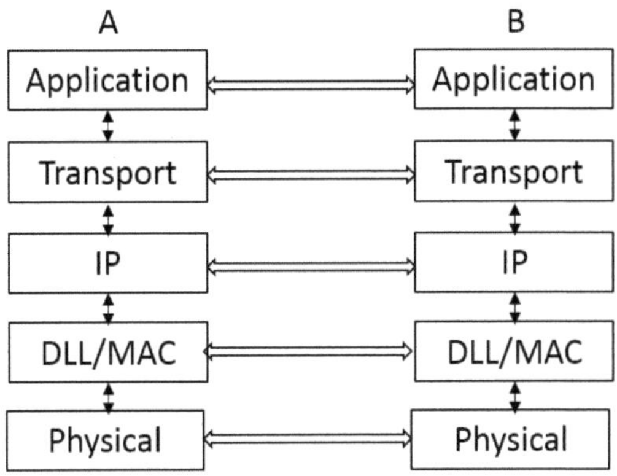

Figure 2.1.2: The Tactical Protocol Stack

6. Media Access Control (MAC) Layer. Used to manage the multiple access media to avoid collisions of data packets from multiple processes or nodes.
7. Physical Layer. Transmits a physical bit sequence over a channel. Each physical channel type uses its own modulation scheme to map the MAC data frames onto the channel. For example, an Ethernet cable modulates data bits as voltages on a twisted pair, a telephone modem maps data bits as tones on a phone line and a radio channel can map data bits to phase changes in a carrier signal.

2.1.2 A Tactical Network Model

The tactical networking community has modified the OSI model into a five layer stack, as shown in figure 2.1.2, centred around the IP protocol as the network layer[1].

The physical layer is layer one, the DLL/MAC layer is layer 2 and the IP layer is layer 3. Within tactical data networks the lower two layers are often combined as the Radio Layer, producing a four layer protocol stack. Note that the interface between the IP layer and the DLL/MAC layer is one-to-many, since a network packet can be transmitted over one of several channels. In an IP enabled radio the device can send IP packets across the radio interface to a distant node, across an Ethernet interface to a local router or across one of the other interfaces supported by the device.

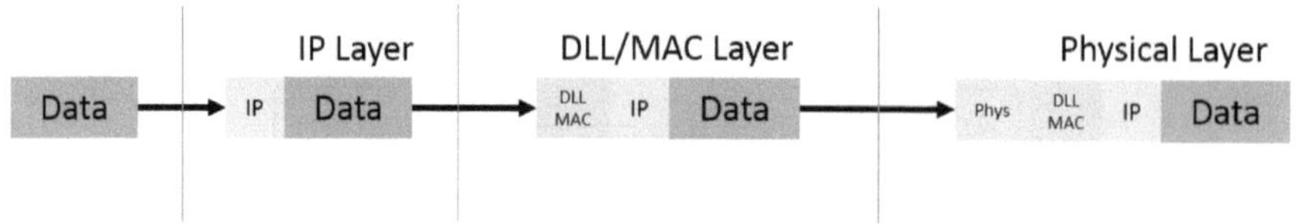

Figure 2.1.3: IP overhead

One of the reasons for the lag in deployment of military IP based radios, when compared to commercial advances, is the amount of overhead required for the IP protocol to function. For each bit of information transmitted over the physical medium there can be a large number of overhead bits to support routing, error correction and other network functions. Since each layer works independently, with its own header information, the ratio of overhead to information can be very high. Each layer generates some header information, but the next layer treats the whole incoming packet or frame as a complete block of data to be encapsulated, as shown in figure 2.1.3. The problem is particularly critical for small payload packets such as Voice Over IP (VoIP). Since tactical networks are generally bandwidth limited the excessive overhead can create very low throughput of information, unless suitable adjustments are made to the performance of each protocol layer.

2.1.3 Security

A significant requirement of the tactical model for the United States and any coalition forces is enforcement of the High Assurance Internet Protocol Encryption (HAIPE) standard at the network layer. HAIPE is defined by the US National Security Agency and is the Type 1 encryption standard for IP-based networks. HAIPE enforces hardware separation of the plain text IP layer and the encrypted IP layer, which creates two network layers at each node. HAIPE effectively allows two classified enclaves to pass data over a lower classified core network. Figure 2.1.4 shows how the OSI layers map onto the lower layers of the tactical network model. HAIPE has very strict rules to protect data from the more secure environment whilst it is transiting through the lower security core network. No routing or topology information is permitted to pass from the more secure enclave to the less secure enclave. However, some packet header elements, such as quality of service, must remain in clear during the transit to ensure that delay sensitive traffic is accorded the required preference from end-to-end within separated secure enclaves.

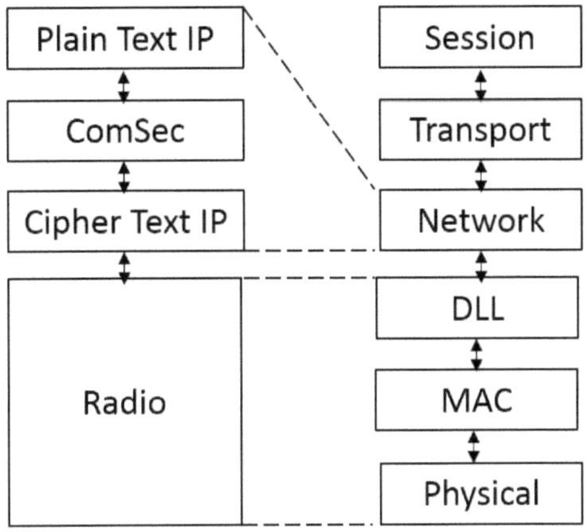

Figure 2.1.4: Mapping the OSI model to secure tactical networks

2.1.4 Digital System Protocol Layer Model

Figure 2.1.5 maps the protocol stack model onto the physical processes necessary in a digital communications system. This diagram is helpful in that it tries to describe the sorts of processes that we should encounter at each stage of the protocol stack. There is not always agreement regarding which layer performs which function, and less so with the introduction of cross-layer signalling, which has been essential to the success of modern tactical data waveforms.

2.2 The Physical Layer

The physical layer in tactical networks defines how a stream of data bits are mapped onto the radio channel. This includes issues such as spectrum usage, modulation schemes, coping with channel noise and signal fading. Implementation of electronic warfare (EW) counter-measures, such as frequency hopping or direct sequence spread spectrum methods, are also implemented in the physical layer.

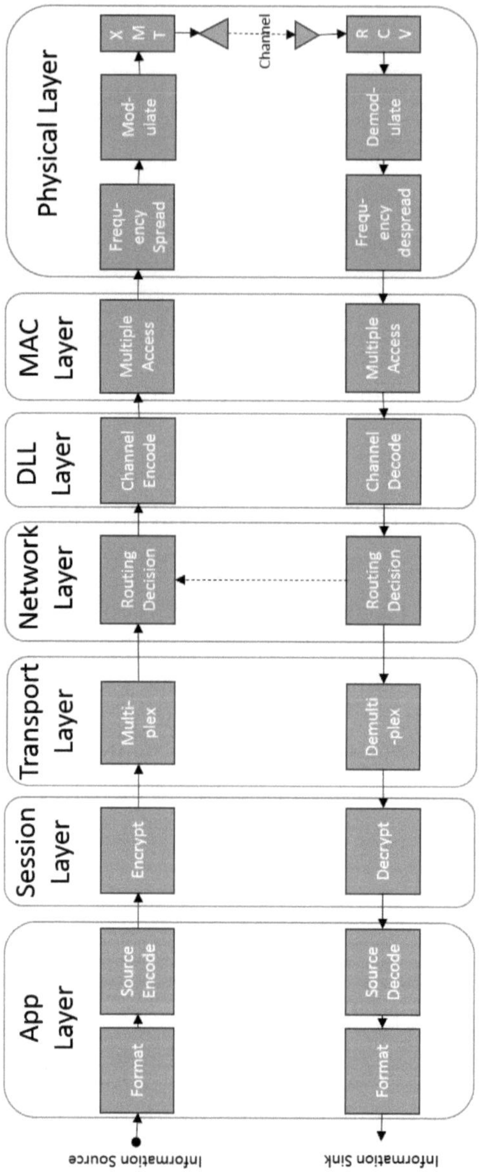

Figure 2.1.5: Digital system protocol layer model

2.2.1 Spectrum Allocation

High Frequency (HF) 3-30 MHz

Beyond line of sight (BLOS) communications remains a challenge for military forces. Although satellite communications is now the preferred BLOS approach, the reality is that satellite channels are a finite resource which may be removed to meet a higher priority force need. Modern improvements to HF technology, particularly Automatic Link Establishment (ALE), has made HF a more convenient and reliable medium. Commercial military HF radios are available that cover the HF/VHF bands from 1.6 MHz - 60 MHz, providing both low speed data and voice communications.

Very High Frequency / Ultra High Frequency (VHF/UHF) 30-300 MHz, 300 MHz - 3 GHz

Military VHF combat net radio has traditionally operated in the 30-70 MHz band and has been used for voice command of tactical force elements. The low VHF band is well suited to operations in undulating terrain, offering greater terrain following properties than the higher frequencies. These radios usually operate with 25 kHz channel spacings.

Modern combat net radios operate across the full VHF spectrum and into the low UHF spectrum. For example, the Harris AN/PRC 152 covers the frequency range from 30-512 MHz, which includes the low UHF satellite band from 240-270 MHz. The AN/PRC 117G covers the frequency range from 30 MHz to 2 GHz, with wider channel bandwidth (up to 5 MHz) available in the higher frequencies.

L Band

L band radios operate in the 1-2 GHz range. Military radios do not generally operate in this band, but the Inmarsat BGAN service does, which is used by many military organisations. The AN/PRC 117G has an advertised capability to interface to certain BGAN terminals in order to establish a range extension facility to the military network.

X Band

X-band is a military satellite band over the 7-8 GHz range. A range of deployable military ground terminals are available for the X-band services, from static terminals with a large dish through to small portable versions. X band offers wideband services to users.

Ka Band

Ka band is used in military satellite communications over the frequencies 30-20 GHz. Ka band offers wideband data services and has been implemented on the US Wideband Global Satellite (WGS) system. The system offers Ka and X band communications with over 4 GHz of bandwidth that can be switched between transponders.

2.2.2 Modulation

Modulation is the particular process of mapping logical bits into a form that is appropriate for the transmission medium. The type of modulation used will vary with the medium: logical values were mapped to audio tones in early telephone modems; logical values are mapped to light and darkness along an optical fibre; logical values are mapped to radiofrequency waves for wireless communications. Radiofrequency mudulation imprints the logical one or zero onto a characteristic of the wave. The three characteristics that can be varied are the amplitude of the wave, the frequency of the wave, or the phase of the wave relative to a standard reference wave. Some modulation schemes use a combination of these characteristics.

Radiofrequency modulation can be broadly classified into two approaches: coherent and non-coherent. Coherent modulation schemes are used when the receiver has access to, or can derive, the same reference timing as that used in the transmitter. Typical coherent digital modulation schemes are Phase Shift Keying (PSK), Frequency Shift Keying (FSK) and Continuous Phase Modulation (CPM), which is also called Minimum Shift Keying (MSK). Non-coherent modulation schemes operate without the use of a transmitter reference signal, and all of the listed schemes have a non-coherent version, except PSK. The closest non-coherent scheme to PSK is Differential PSK, which, in the absence of a reference, uses the phase of the previously received bit to determine the phase of the currently received bit. Generally, the performance of the non-coherent schemes in the presence of noise is slightly worse than their coherent counterparts. Unfortunately, in tactical networks non-coherent modulation schemes are usually required due to the ad-hoc nature of the networks [6].

An ideal modulation scheme would map the logical bits in such a way that when we demodulate the received signal at the receiver we can be sure that we have mapped it to the correct sequence of transmitted logical bits. The scheme would maximise the data rate; minimise the probability of bit error; minimise the transmission power required; minimise the bandwidth required; and minimise the complexity of the electronic circuit. Unfortunately, no scheme does all of these things well all the time, so the choice of coding scheme needs to take into account the communication channel that is going to be used. Channels always introduce distortion, so the selected scheme should be robust against the type of distortion likely to be encountered in the chosen communications channel. This inevitably leads to a trade-off of capacity for error performance, or power required against bandwidth.

Orthogonal Signals in Space

Modulation maps the digital data stream onto a carrier wave for radio transmission, and the digital data is imprinted on the wave by modifying either the frequency, the amplitude or the phase. The transmitted waveforms are referred to as symbols and for binary schemes we are mapping one bit per symbol. For example, in binary FSK the two symbols are two different carrier frequencies, one representing a digital one and the other representing a digital zero. For binary PSK the two symbols are are the two phases of the carrier frequency, typically zero (or in-phase) and π radians (or "out of phase"). Irrespective of the particular scheme being used, we can refer to these symbol waveforms as basis functions, ϕ_1 and ϕ_2, and at any instant in time we are transmitting a signal, $s(t)$, which is a linear combination of these basis functions, as illustrated in figure **2.2.1**. To aid detection it preferable for the basis functions in each scheme to be orthogonal.

$$s(t) = c_1\phi_1(t) + c_2\phi_2(t).$$

For the binary PSK case it turns out that we can use only a single basis function, say ϕ_1, and c_1 alternates between ± 1, which is called antipodal signalling. For more sophisticated modulation schemes it is not quite so simple.

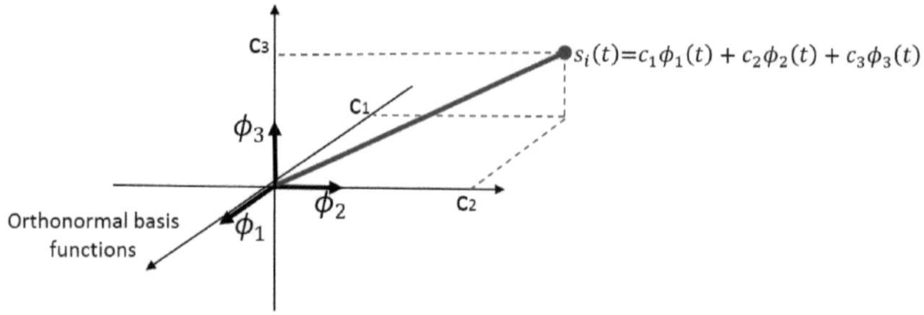

Figure 2.2.1: Orthogonal signal components

PSK Example

Assume that we need a PSK scheme in which we modulate the signal for a period of T_s seconds, while using a reference signal energy equal to E_s Joules. If we choose

$$\begin{aligned} \phi_1(t) &= \left(\frac{2}{T_s}\right)^{\frac{1}{2}} \cos(\omega_c t), \\ \phi_2(t) &= \left(\frac{2}{T_s}\right)^{\frac{1}{2}} \sin(\omega_c t) \end{aligned}$$

then our basis functions have the useful property of being orthonormal, or

$$\int_0^{T_s} \phi_i \phi_j dt = \delta_{ij} = \begin{cases} 1, & i = j \\ 0, & i \neq j \end{cases}$$

which will be become important during signal detection. Since the two basis functions are orthogonal, they span a two dimensional vector space and hence, for a given carrier frequency ω_c we can transmit a signal of arbitrary phase by using a linear combination of ϕ_1 and ϕ_2[1].

Example. A signal that is a linear combination of ϕ_1 and ϕ_2 has the general form $s(t) = \left(\frac{2E_s}{T_s}\right)^{\frac{1}{2}} [A\cos(\omega t) + B\sin(\omega t)]$ where A and B are constants and E_s is the average symbol energy. Choose A and B such that $A = \cos(\alpha)$

[1]The basis functions are orthonormal if the carrier frequency, f_c, is an integer multiple of the symbol rate R_b. However, even when this criterion is not met the effect is negligible if $f_c \gg R_b$. See appendix 4 for an explanation of why this is so.

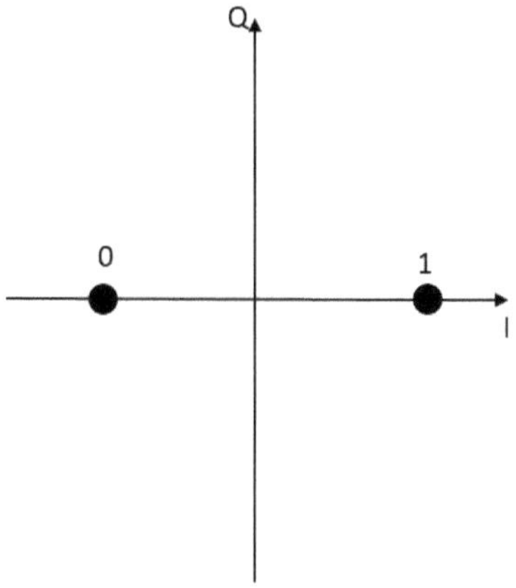

Figure 2.2.2: BPSK constellation diagram

and $B = \sin(\alpha)$ for some desired phase angle α. Then

$$\begin{aligned} s(t) &= \left(\tfrac{2E_s}{T_s}\right)^{\frac{1}{2}} [\cos(\omega t)\cos(\alpha) + \sin(\omega t)\sin(\alpha)] \\ &= \left(\frac{2E_s}{T_s}\right)^{\frac{1}{2}} \cos(\omega t - \alpha) \end{aligned}$$

which is a signal with phase α.

Now, for the binary case, we want $s_1(t) = \left(\tfrac{2E_s}{T_s}\right)^{\frac{1}{2}} \cos(\omega_c t)$ (in-phase) and

$$\begin{aligned} s_2(t) &= \left(\frac{2E_s}{T_s}\right)^{\frac{1}{2}} \cos(\omega_c t + \pi) \\ &= -\left(\frac{2E_s}{T_s}\right)^{\frac{1}{2}} \cos(\omega_c t) \end{aligned}$$

(out-of-phase) which means that the binary signal can be represented in a single dimension using only $\phi_1(t)$. This scheme is usually drawn as a constellation diagram as shown in figure 2.2.2 where the I-axis represents ϕ_1 and the Q-axis represents ϕ_2.

,

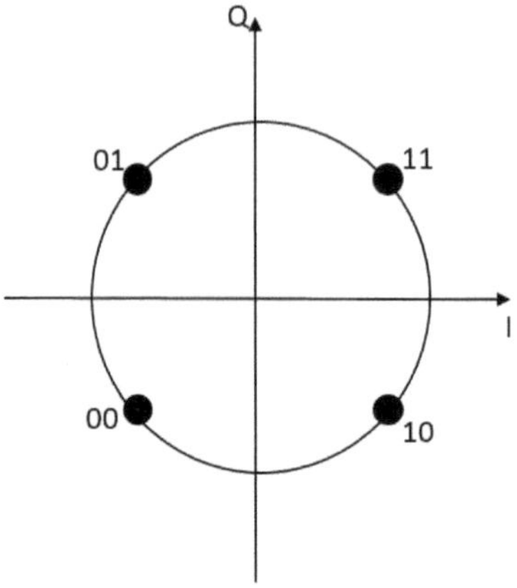

Figure 2.2.3: Cartesian representation of QPSK

M-ary PSK

In more complex modulation schemes when each symbol is transmitted it carries more than one bit of information. Quadrature PSK (QPSK) transmits four phases instead of two and hence can allocate two bits per symbol, since there are four unique combinations of two bits: 00, 01, 10 or 11. This is an advantage because for the same symbol rate we can double the underlying bit rate, and since the bandwidth requirement is determined by the symbol rate we effectively double the information we can send down the same channel bandwidth.

In the QPSK case we have, ignoring scale factors, four possible signals to be transmitted: $\pm\cos(\omega_c t)$ and $\pm\sin(\omega_c t)$, or more usually

$$s_i(t) = \left(\frac{2E_s}{T_s}\right)^{\frac{1}{2}} \cos\left(\omega_c t + i\frac{\pi}{2}\right),\ i = 0, 1, 2, 3.$$

The term quadrature is used because each of the four signals are separated by $\frac{\pi}{2}$ radians of phase. The constellation diagram is shown in figure 2.2.3.

QPSK is commonly available in tactical radios and can be generated as illustrated in figure 2.2.4. Because the cosine and sine functions are orthogonal, each can be modulated separately, as a binary PSK system, and the two binary signals can then be summed to form the quadrature signal. One signal is called the I signal (In-phase) and the other is called the Q signal (Quadrature). Practically, a single carrier, generated by a local oscillator (L.O.) circuit

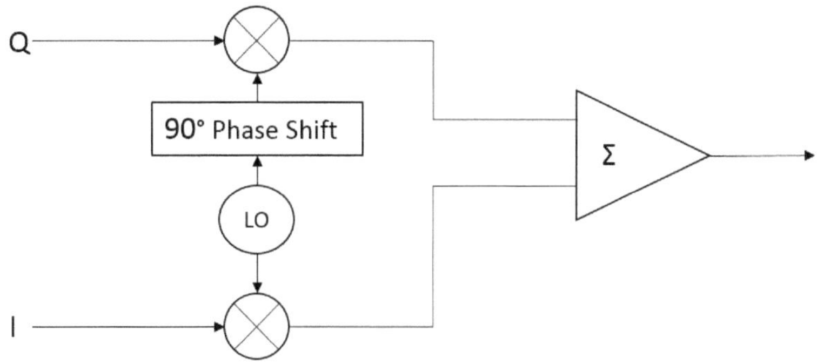

Figure 2.2.4: QPSK Generation

is split into two paths. One path is delayed by an amount of time equal to one quarter of the carrier's cycle time, or 90 degrees of phase, to create the sine wave. The two carriers are amplitude modulated, one by the I signal, the other by the Q signal and the two modulated carriers are added together in a summing circuit. The output is a digitally modulated signal whose amplitude and phase are determined by the amplitudes of the two modulating signals.

Leaving scale factors aside, the four possible outputs from this circuit are:

- $\sin(\omega_t) + \cos(\omega_t)$
- $\sin(\omega_t) - \cos(\omega_t)$
- $-\sin(\omega_t) + \cos(\omega_t)$
- $-\sin(\omega_t) - \cos(\omega_t)$

which produces four cosine waves with phases of 45°, 135°, 225° or 315°, as illustrated in figure 2.2.3.

It is possible to construct even more complex schemes, such as 8-PSK, which can transmit three bits per symbol, or to include amplitude variation in the modulation to produce 16-QAM (Quadrature Amplitude Modulation) which transmits four bits per symbol. The constellation diagram for 16-QAM is shown in figure 2.2.5, which illustrates how much closer together the constellation points are than QPSK. This is significant because in the presence of noise we need to make a decision about which symbol has been received. Clearly, if the constellation points are close together then when a signal is corrupted by noise it becomes difficult to determine which constellation point represents the true signal and wrong decisions increase the bit error rate. Unfortunately, the tactical communications environment is usually a noisy environment, so correctly detecting signals is always a trade off between power,

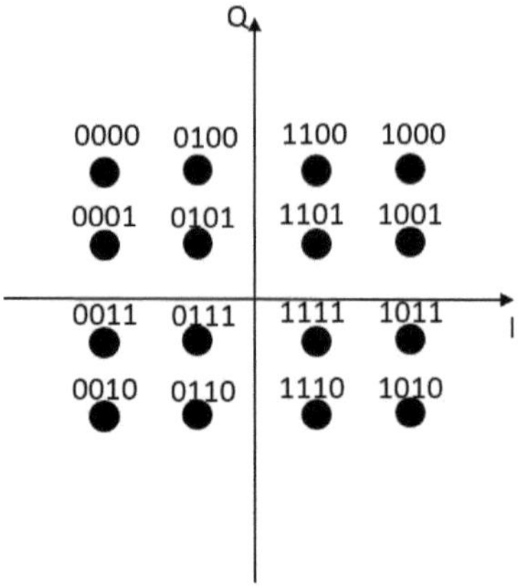

Figure 2.2.5: 16-QAM Constellation

bandwidth and complexity.

2.2.3 Demodulation and Detection in Noise

Noise in a tactical radio channel is usually modelled as Additive White Gaussian Noise (AWGN), not least because it makes the calculation of error probabilities tractable. The addition of noise to a transmission means that when a corrupted symbol arrives a decision must be made regarding which source symbol was actually transmitted. Since the spectrum of AWGN is flat it is characterised by the noise power spectral density, N_0, which is assumed constant across all frequencies. Hence, to calculate noise power (N or σ_0^2) we need to multiply N_0 by the appropriate bandwidth. In order to compare the probability of error in a noisy channel for different modulation schemes the common metric used is $\frac{E_b}{N_0}$, where E_b is the average energy available per bit.

Figure 2.2.6 shows the known constellation of an 8-ary modulation scheme in which there are eight known alphabet symbols. A noisy distorted signal has been received which does not exactly match any of the eight valid symbols, so a decision must be made about which symbol was *most likely* to have been transmitted. The decision is made by measuring the distance between the received symbol and each of the possible transmitted symbols. Whichever of the known symbols that the received symbol is closest to is assumed to be the most likely transmitted symbol.

The practical method of measuring that distance is by using a matched fil-

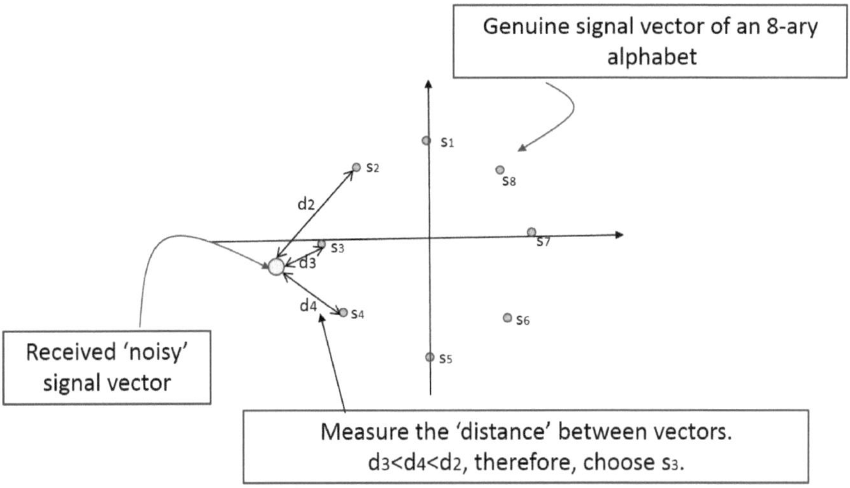

Figure 2.2.6: Detection in noise

ter arrangement, which can correlate the received signal with replicas of the known symbol alphabet. For example, if BPSK is being used then the matched filter will correlate the received signal against $\cos(\omega_c t)$ and $\cos(\omega_c t + \pi)$. The decision is made by assuming that the largest correlation, or shortest distance, will be produced by the symbol replica that matches the actual symbol transmitted. Each modulation scheme has advantages and disadvantages, with the probability of bit error being a significant metric.

Coherent and Non-Coherent Detection

Coherent detection relies on the receiver having access to a reference signal that represents the phase of the transmitter. If such a signal is available then coherent detection schemes can be employed, which can sometimes improve error performance. However, a high end oscillator at 200 MHz can shift by 1 Hz per second, which is quite small theoretically but still requires some form of phase locking. In the tactical domain node mobility can also cause the phase of the carrier signal to shift, due to the Doppler effect, and channel fading is also a constraint on maintaining phase lock. Constant phase locking periods become very inefficient, particularly in channels that transmit in bursts, such as time division multiple access (TDMA). In tactical communications non-coherent binary schemes have been favoured for their robustness,

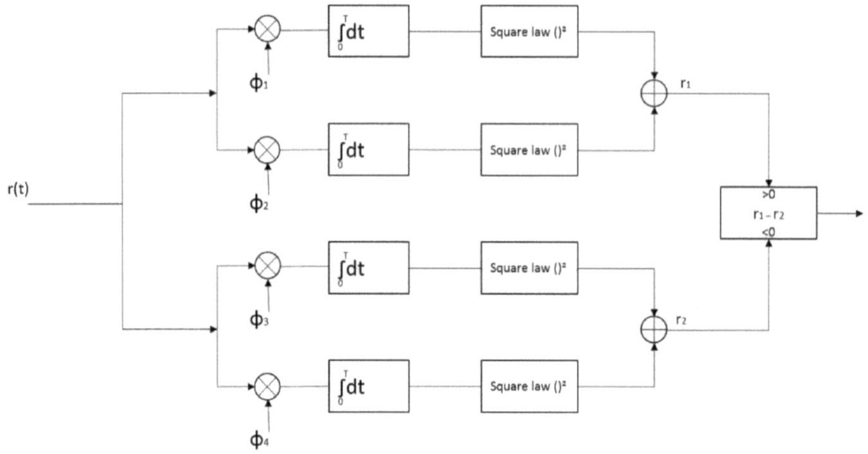

Figure 2.2.7: Non-coherent BFSK Receiver

with binary FSK being the original tactical data modulation scheme.

Error Performance: Non-Coherent BFSK

The SINCGARS waveform uses binary FSK for the transmission of data. That is, a digital one is represented by frequency ω_0 and a digital zero is represented by a symbol transmitted at frequency ω_1. For coherent detection of BFSK a two element correlator can be used. However, tactical radios do not usually have the luxury of coherency due to volatility of the radio transmission path through fading or terrain shielding or the effects of mobile operations on radio transmissions. Non-coherent detection of BFSK can still be achieved through a correlator arrangement, but since the phase of the received signal is unknown the filter requires four branches rather than two. The non-coherent detector correlates the I and Q channels of the received signal and then sums the results to determine the overall correlation of the received signal with each possible source symbol. A non-coherent BFSK correlator detector is illustrated in figure 2.2.7.

For the purpose of illustrating the relationship between bit error rate and energy per bit, an alternate demodulation method will be used. Figure 2.2.8 shows a matched filter detector in which two filters are used, one centred at frequency f_0 and the other at frequency f_1. The bandpass of the filters is $W_f = \frac{1}{T}$ and the filtered received signals are then passed through an envelope

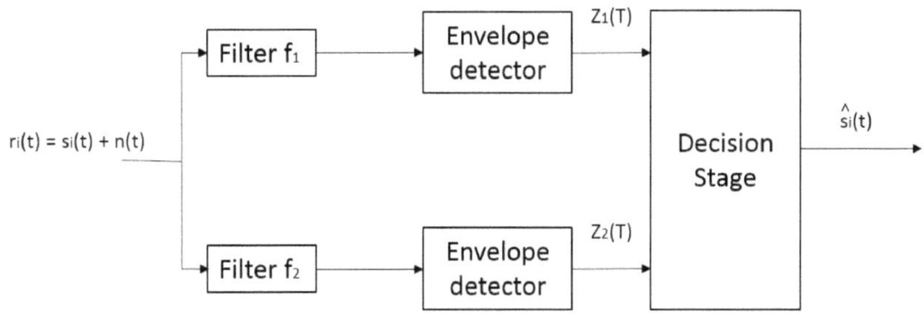

Figure 2.2.8: Non-coherent FSK Matched Filter Detector

detector. The envelope detectors consist of a rectifier and a low pass filter and they are matched to the signal envelopes, not the signals themselves. Therefore, no phase information is required. The decision is made on the basis of which envelope detector has the largest value at the moment of decision.

For the matched filter detector to be effective the signal set must comprise orthogonal signals. For non-coherent BFSK it can be shown [6] that the minimum frequency separation required to achieve an orthogonal signal set is $\Delta f = \frac{1}{T}$, where Δf is the frequency separation and T is the symbol period (bit period for a binary system). Consider the equally likely binary signal set $\{s_i(t)\}$, defined as

$$s_i(t) = \sqrt{\frac{2E_b}{T}} \cos\left(\omega_i t + \phi\right) \quad 0 \leq t \leq T, \quad i = 1, 2$$

where the phase term ϕ is unknown and assumed constant, and E_b represents the energy per bit transmitted. The addition of AWGN to the transmitted signal means that the decision statistic $z(T)$,in figure 2.2.8, has a Gaussian distribution as shown in figure 2.2.9. Intuitively, we can see that the probability of a bit error is the area of the tail on the "wrong side" of the decision line.

The input to the detector in figure 2.2.8 is $r(t) = s_i(t) + n(t)$, where $n(t)$ is a white Gaussian noise process with two sided power spectral density $\frac{N_0}{2}$. Let H_1 be the hypothesis at the receiver that signal s_1 was transmitted and H_2 be the hypothesis that signal s_2 was transmitted. If we assume that both signals

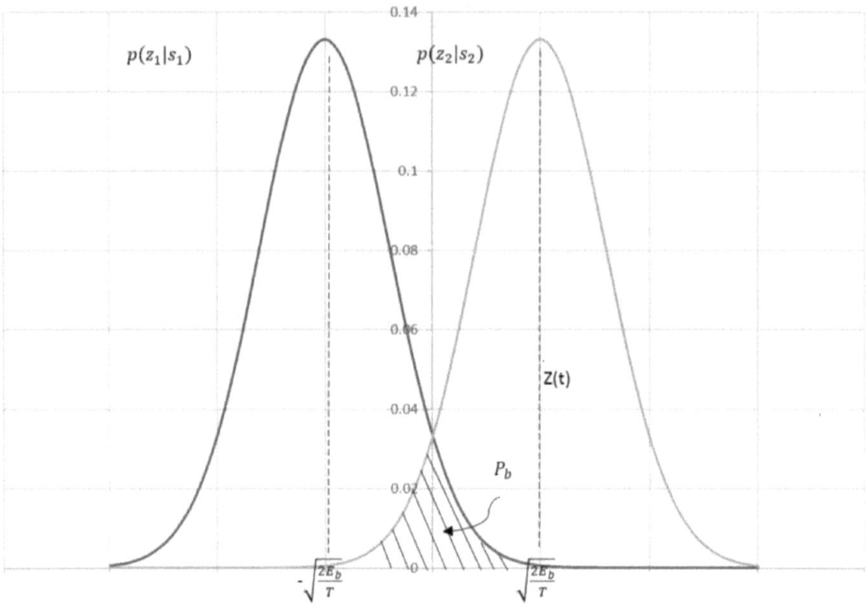

Figure 2.2.9: Conditional Probability Density Functions: $p(z_1|s_1), \quad p(z_2|s_2)$

are equally likely then the probability of bit error can be expressed as

$$\begin{aligned} P_B &= \frac{1}{2}P\left(H_2|s_1\right)+\frac{1}{2}P\left(H_1|s_2\right) \\ &= \frac{1}{2}\int_{-\infty}^{0} p(z_2|s_1)dz+\frac{1}{2}\int_{0}^{\infty} p(z_1|s_2)dz \end{aligned}$$

The test statistic $z(T) = z_1(T) - z_2(T)$. In a noise free environment the value of $z(T) = \sqrt{\frac{2E_b}{T}}$ when $s_1(t)$ is sent and $z(T) = -\sqrt{\frac{2E_b}{T}}$ when $s_2(t)$ is sent. Therefore, by symmetry

$$\begin{aligned} P_B &= \int_{0}^{\infty} p(z_1|s_2)dz \\ &= P\left(z_1 > z_2|s_2\right) \end{aligned}$$

In the case when $s_2(t)$ is sent, the output of the filter f_1 is a Gaussian noise random variable only, since no signal component is present. A Gaussian distribution into the non-linear envelope detector produces a Rayleigh distribution at the output [8], so that

$$p(z_1|s_2) = \begin{cases} \frac{z_1}{\sigma_0^2} \exp\left(-\frac{z_1^2}{2\sigma_0^2}\right) & z_1 > 0 \\ 0 & z_1 < 0 \end{cases}$$

where σ_0^2 is the noise power at the filter output. In the other leg of the filter, $z_2(t)$ has a Rician distribution [8], since the input to the lower envelope detector is a sinusoid plus noise. Hence, the probability density function, $p(z_2|s_2)$, is written as

$$p(z_2|s_2) = \begin{cases} \frac{z_2}{\sigma_0^2} \exp\left(-\frac{(z_2^2+A^2)}{2\sigma_0^2}\right) I_o\left(\frac{z_2 A}{\sigma_0^2}\right) & z_2 \geq 0 \\ 0 & z_2 < 0 \end{cases}$$

where $A = \sqrt{\frac{2E_b}{T}}$. [2] When $s_2(t)$is transmitted, the receiver makes an error when the envelope sample $z_1(t)$ obtained from the upper channel exceeds the sample $z_2(t)$ obtained from the lower channel. The probability of this error can be found by integrating $p(z_1|s_2)$ with respect to z_1 from z_2 to infinity, then averaging over all possible values of z_2.

$$\begin{aligned} P_B &= P(z_1 > z_2|s_2) \\ &= \int_0^\infty p(z_2|s_2)\left[\int_{z_2}^\infty p(z_1|s_2)dz_1\right]dz_2 \\ &= \int_0^\infty \frac{z_2}{\sigma_0^2}\exp\left(-\frac{(z_2^2+A^2)}{2\sigma_0^2}\right) I_o\left(\frac{z_2 A}{\sigma_0^2}\right)\left[\int_{z_2}^\infty \frac{z_1}{\sigma_0^2}\exp\left(-\frac{z_1^2}{2\sigma_0^2}\right)dz_1\right]dz_2 \end{aligned}$$

where $A = \sqrt{\frac{2E}{T}}$. This integral can be evaluated to yield

$$\begin{aligned} P_B &= \frac{1}{2}\exp\left(-\frac{A^2}{4\sigma_0^2}\right) \\ &= \frac{1}{2}\exp\left(-\frac{A^2}{4N_0 W_f}\right) \\ &= \frac{1}{2}\exp\left(-\frac{E_b}{2N_0}\right) \end{aligned} \qquad (2.2.1)$$

[2]The function $I_o(x)$ is a modified zero-order Bessel function of the first kind, defined as $I_o(x) = \frac{1}{2\pi}\int_0^{2\pi}\exp(x\cos\theta)d\theta$.

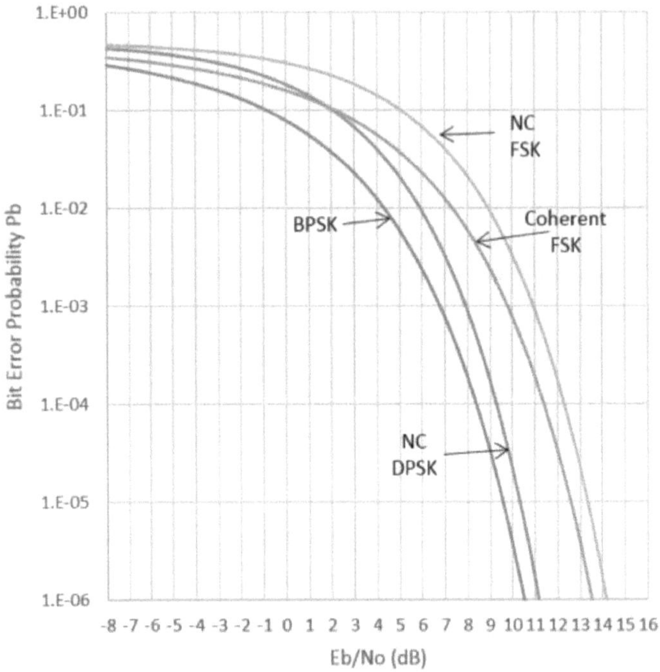

Figure 2.2.10: Bit error probability for common digital modulation schemes

where N_0 is the noise power spectral density and $W_f = \frac{1}{T}$ is the filter bandwidth. A plot of the function in equation 2.2.1 produces the classic waterfall graph for digital communications. Figure 2.2.10 compares the bit error performance for several coherent and non-coherent modulation schemes and table 2.1 lists the P_B expression for other common schemes.

Note the presence of $\frac{E_b}{N_0}$ in each expression in table 2.1. This ratio is the commonly used measure of error performance in digital communications, rather than signal-to-noise ($\frac{S}{N}$), because it measures the underlying bit error rate after demodulation. However, $\frac{E_b}{N_0}$ is related to $\frac{S}{N}$ by the following relation

$$\frac{E_b}{N_0} = \frac{ST}{N_0} = \frac{S}{RN_0} = \frac{SW}{RN_0W} = \frac{S}{N}\left(\frac{W}{R}\right) \tag{2.2.2}$$

Modulation	P_B
Coherent PSK	$Q\left(\sqrt{\frac{2E_b}{N_0}}\right)$
Non-coherent DPSK	$\frac{1}{2}\exp\left(-\frac{E_b}{N_0}\right)$
Coherent FSK	$Q\left(\sqrt{\frac{E_b}{N_0}}\right)$
Non-coherent FSK	$\frac{1}{2}\exp\left(-\frac{1}{2}\frac{E_b}{N_0}\right)$

Table 2.1: Bit error performance of common modulation schemes

where

$$\begin{aligned} S &= \text{average modulating signal power} \\ T &= \text{bit time duration} \\ R &= \text{bit rate} = \frac{1}{T} \\ N &= N_0 W. \end{aligned}$$

2.2.4 Fading

Fading is the temporary loss of communications due to channel variation. Fading effects include signal amplitude distortion and also signal phase distortion.

Deep signal amplitude fading can occur in tactical radio channels which can severely degrade communications for short bursts of time. Amplitude fading can be modelled by a channel gain, $A(t)$ that varies randomly in time as a Rayleigh random process. Rayleigh models are used when multi-path reflected signals are present, creating a destructive interference effect at the receiver, as shown in figure 2.2.11. When the main path transmission is present the amplitude variation is typically modelled as a Rician distribution. Modulation schemes that depend on amplitude measurements, such as QAM, should be avoided in fading channels because there will be significant amplitude variation in the received signal.

Doppler effects due to mobility, as illustrated in figure 2.2.12, create phase uncertainty in tactical communication systems. Some of this uncertainty can be compensated for by increasing the available power, which improves the $\frac{E_b}{N_0}$ ratio, but Doppler fading, which spreads the transmission in the frequency domain, can still cause significant deep fading effects. Non-coherent schemes are essential because this type of fading also creates phase distortion. Figure

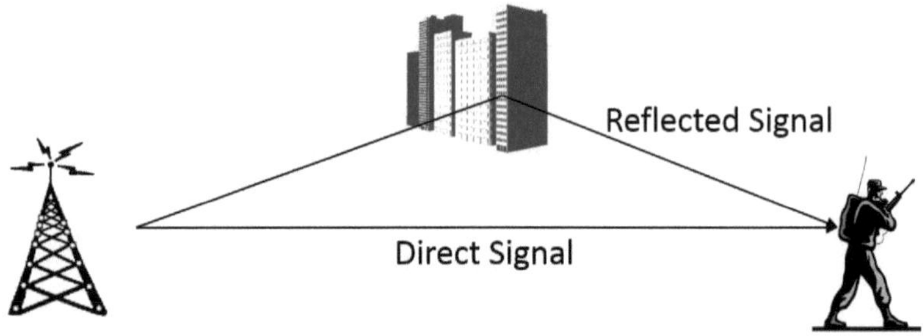

Figure 2.2.11: Multi-path amplitude fading

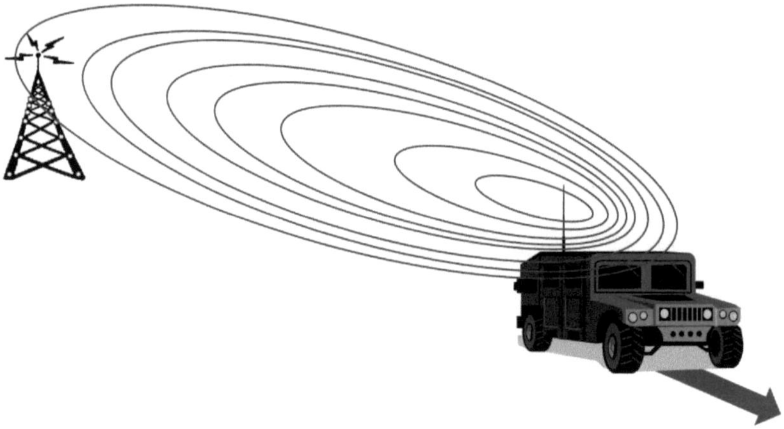

Figure 2.2.12: Doppler fading

2.2.13 is an example of the deep amplitude fading effects that can be experienced in tactical radio channels.

Our aim is to have some way to model the effects of fading so that we can compensate for it in the design of our system. For example, assume we are using coherent BPSK as our modulation scheme and we need a bit error performance of $P_B = 10^{-5}$, then we need $\frac{E_b}{N_0} = 9.6$ dB =9.12. If we assume the required bit rate is 32 kbps and the channel is a standard 25 kHz, then using equation 2.2.2 we need a SNR of

$$\frac{S}{N} = \frac{R}{W}\frac{E_b}{N_0} = \frac{32,000}{25,000} \cdot 9.12 = 11.6 = 10.7 \text{ dB}.$$

On top of this we also need a suitable power margin to compensate for the fading effects. To calculate this margin we start by trying to estimate the

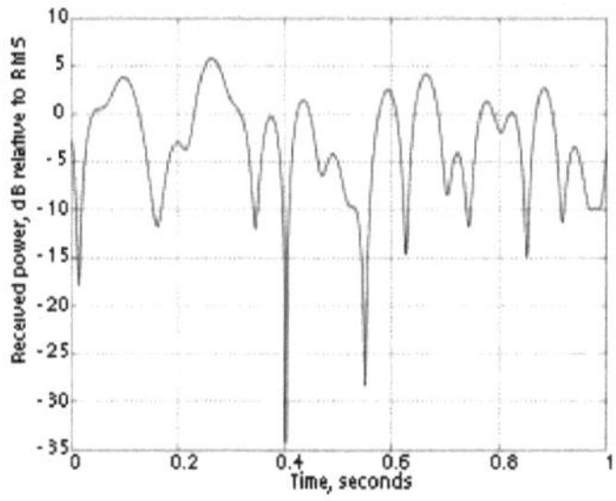

Figure 2.2.13: Example Rayleigh Fading Channel Response

average error probability over a fading period.

If the channel had a fixed gain, $a \geq 0$, then the probability of error can be expressed as $P(\epsilon|a)$, where the ϵ event is the occurrence of a symbol error. However, fading is a Rayleigh random process, expressed as $f_A(a) = 2ae^{-a^2}$, so the average error probability is calculated as [1]

$$P_s = \int_0^\infty P(\epsilon|a) f_A(a) da. \tag{2.2.3}$$

In our example BPSK has, from table 2.1, a probability of bit error given by $P_B = Q\left(\sqrt{\frac{2E_b}{N_0}}\right)$. In the presence of Rayleigh fading, if the channel amplitude is a, then the conditional probability of bit error (bit and symbol error are the same for binary schemes) is given by

$$P(\epsilon|a) = Q\left(\sqrt{\frac{a^2 2E_b}{N_0}}\right). \tag{2.2.4}$$

Inserting equation 2.2.4 and the Rayleigh expression into equation 2.2.3 gives

$$P_B = \int_0^\infty Q\left(\sqrt{\frac{a^2 2E_b}{N_0}}\right) 2ae^{-a^2} da$$

which can be evaluated as

$$P_B = \frac{1}{2}\left(1 - \left[\frac{\frac{E_b}{N_0}}{1 + \frac{E_b}{N_0}}\right]^{\frac{1}{2}}\right). \tag{2.2.5}$$

For $\frac{E_b}{N_0} \geq 13\text{dB}$, equation 2.2.5 can be approximated as [1]

$$P_B \approx \frac{1}{4E_b/N_0} \tag{2.2.6}$$

with a variation from the exact figure of less than 4%.

What do we conclude from equation 2.2.6? In order to maintain a $P_B = 10^{-5}$ with a coherent BPSK modulation scheme in the presence of Rayleigh fading we need $\frac{E_b}{N_0} = 44$ dB, which is an enormous increase (2500 times) on the 9.6 dB required in a non-fading environment. Clearly it is impractical to increase the power in every transmitter in order to compensate for this effect, but there may be alternate ways to provide an acceptable performance level.

Firstly, this is the average probability of bit error in the channel in the presence of fading. Yet fading events are usually quite short in duration, so satisfactory (10^{-5}) performance can be achieved during the non-fading periods without increasing the transmit power. An acceptable performance during a fade event might be $P_B = 10^{-4}$, which would require an increase in the transmit power of only 10 dB, which is more manageable. Alternately, it may be acceptable to require $P_B = 10^{-5}$ for 90% of the time which requires no increase in SNR, since the fade events are short. Instead, error control coding can be used to improve error performance during the fade event, although this also introduces overhead bits which reduce throughput. As with all engineering, there is no such thing as a free lunch.

A similar analysis on coherent and non-coherent orthogonal signals, such as FSK, with Rayleigh fading produces an equivalent to equation 2.2.6 of

$$P_B \approx \frac{1}{2E_b/N_0} \text{ and } P_B \approx \frac{1}{2 + E_b/N_0}$$

respectively, which shows that non-coherent signals perform poorer than coherent signals by a factor of 4-6 dB.

2.2.5 Power Spectrum

Analysing the power spectrum of a modulation scheme can provide useful information about the effectiveness of the scheme under different channel conditions. A power spectrum is an averaged view of the distribution of energy in a signal across the bandwidth. The period of averaging can be important, since the longer the averaging occurs the smoother the spectrum appears, which can

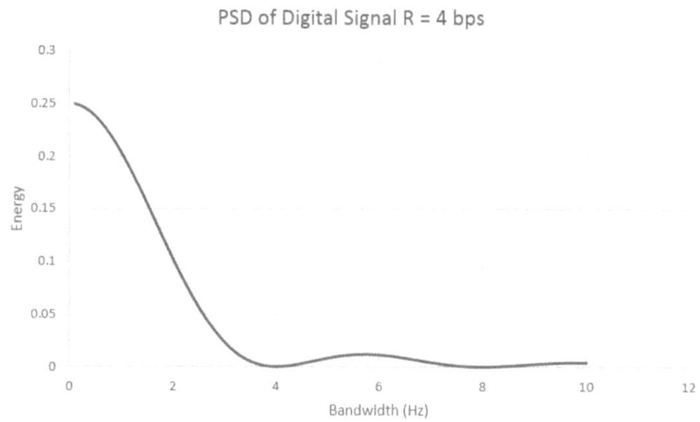

Figure 2.2.14: Power Spectral Density of a Baseband Digital Signal

mask some critical signal features.

Each modulation scheme produces a different power spectrum, which should be considered when selecting a waveform for tactical communications.

Baseband Spectrum

The spectrum of a 4 bit per second baseband digital signal is shown in figure 2.2.14. The low bit rate has been chosen for illustration purposes. A digital signal has an infinite bandwidth, but most of the energy is contained in the central lobe which has a width of $f = \frac{1}{T_b}$ where T_b is the bit period. That is, for a baseband digital signal the width of the main lobe has the same value in Hertz as the bit rate of the signal. A modulated signal is often band limited by the channel, so the power spectrum is truncated, but in such a way that the distortion introduced by removing some frequencies does not adversely affect the error performance.

M-PSK

Since PSK is generated by multiplying a sinusoid by the baseband digital signal, the power spectrum of PSK modulation is the convolution of the two-sided version of the spectrum in figure 2.2.14 with the single spike of the sinusoid at the carrier frequency. This process produces the spectrum shown in figure 2.2.15, where the main lobe contains most of the energy of the PSK

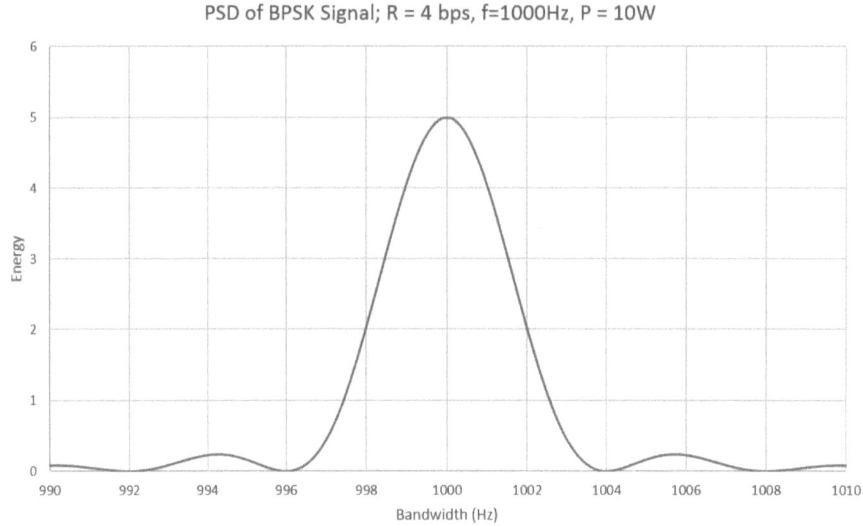

Figure 2.2.15: Power Spectral Density of M-PSK signals

signal and is typically the portion transmitted. In the time domain the PSK signal is illustrated in figure 2.2.16. Note the sharp phase transitions that align with each bit transition. It is the sharp transitions in the time domain that create the distinctively wide tails on the spectrum in the frequency domain.

For M-PSK systems, adding additional phase constellation points does not change the spectrum of the modulated signal. Therefore, M-PSK is useful in band limited channels because additional bits per symbol can be transmitted within the same bandwidth, increasing the data rate, or the same data rate can be transmitted in less bandwidth. Of course, cramming the constellation phase points closer together also increases the probability of bit error, so this must also be considered. Note that for BPSK with a data rate of R bits per second the first *null* in the spectrum occurs at R Hertz above and below the carrier frequency. This is also a common measure for the bandwidth of the signal, since the small side lobes can be suppressed without creating too much distortion in the signal.

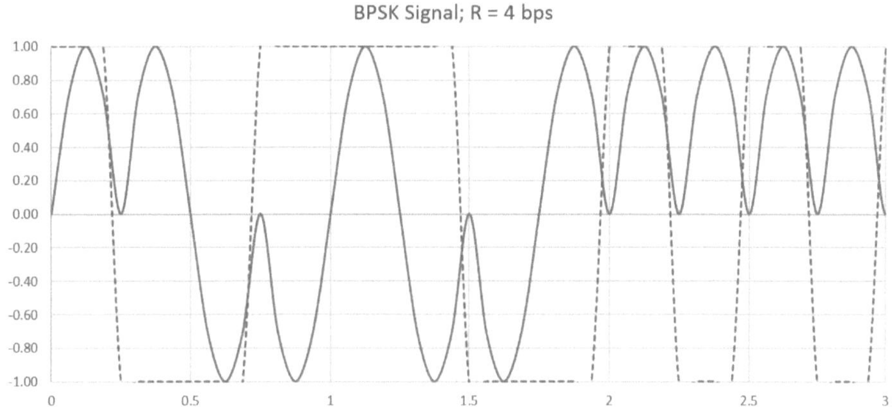

Figure 2.2.16: PSK time domain signal

Continuous Phase Modulation (CPM)

Continuous phase modulation (CPM) is a variation on PSK that is offered in some commercial tactical radios, such as the Harris AN/PRC 152 VHF/UHF radio. The side lobes in PSK transmissions are created by the sharp changes in signal voltage that occur at the phase transition points. CPM modulates the data by changing phase, but makes the change slowly throughout the bit period by using a sinusoidal weighting. Avoiding sharp transitions reduces the spectrum "splash", which results in less out-of-band energy. The end result is more energy in the pass band, which improves the SNR. Figure 2.2.17 compares the spectrum of BPSK and QPSK with CPM for the same data rate. Note the reduced side lobes for CPM.

FSK

Orthogonal FSK can be imagined as two carriers, each modulated by its own data stream. The produced power spectrum, as shown in figure 2.2.18, looks like two separate PSK spectrum, although usually quite close together in order to maintain spectral efficiency. Of course, we do not want the spectrum from one carrier to interfere with the spectrum of the other carrier, so the spacing between carrier frequencies is important. In order to achieve orthogonality, and hence improve signal detection, the spacing between the frequencies, f_1 and f_2, should be a minimum of $\frac{1}{T}$ for non-coherent detectors and $\frac{1}{2T}$ for

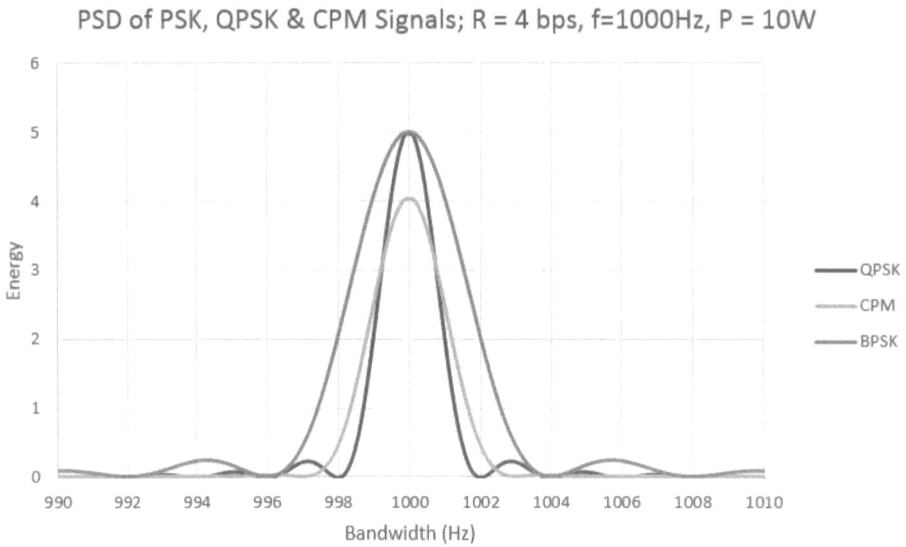

Figure 2.2.17: Spectrum of CPM and BPSK

coherent detectors. This ensures that, on average, when the receiver is trying to detect which basis signal is present in the received signal it is not receiving spurious energy from the symbol lag of the other carrier. Symbol lag occurs because the spectrum has been band limited through the channel.

2.2.6 Minimum Transmission Bandwidth

The power spectra images for PSK and FSK show that the frequency spectrum of a signal modulated in this way has an infinite set of frequency components. For example, although most of the energy in a PSK signal is contained in the main lobe, there is some energy spread across all frequencies, as indicated by the ever reducing nulls in the spectrum shape. However, real signals are always transmitted through a band limited channel, usually due to regulation and the need to share the radio frequency spectrum among many users. By limiting the bandwidth of the channel some energy must be lost from the original modulated signal, which creates distortion of the signal at the receiver, but it is instructive to consider exactly what type of distortion occurs.

Intuitively, one might expect fast changing signals to have high frequency components, and that is the case. Sharp transitions in the time domain require

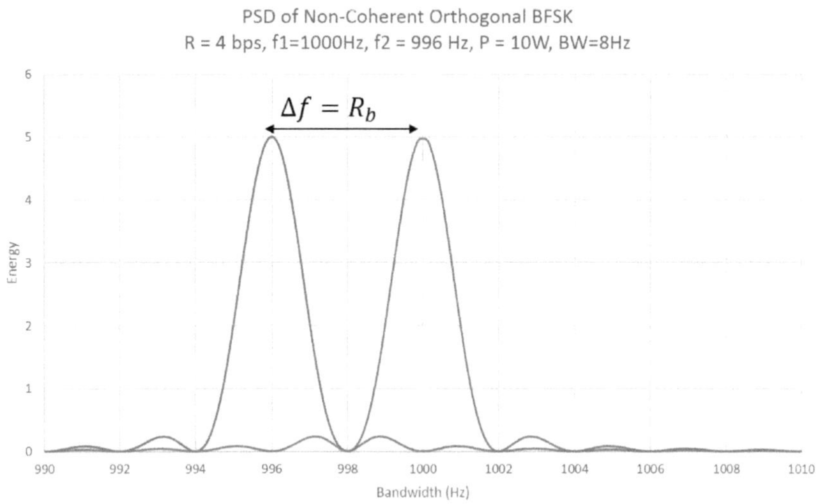

Figure 2.2.18: Power Spectral Density of non-coherent BFSK

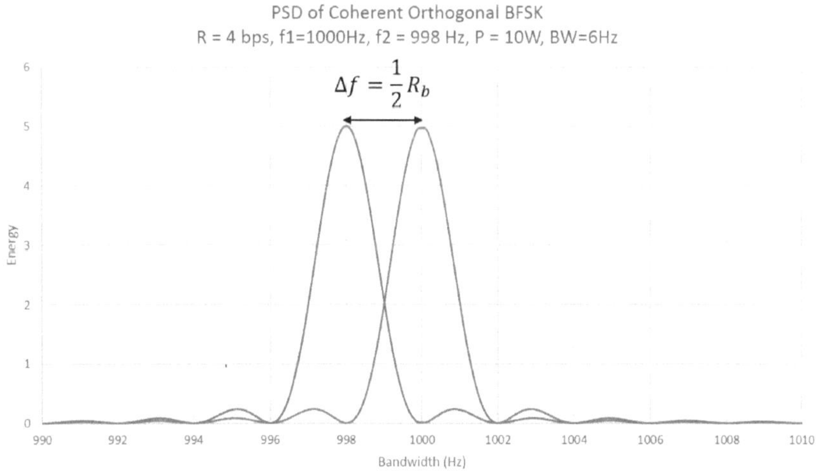

Figure 2.2.19: Power Spectral Density of coherent BFSK

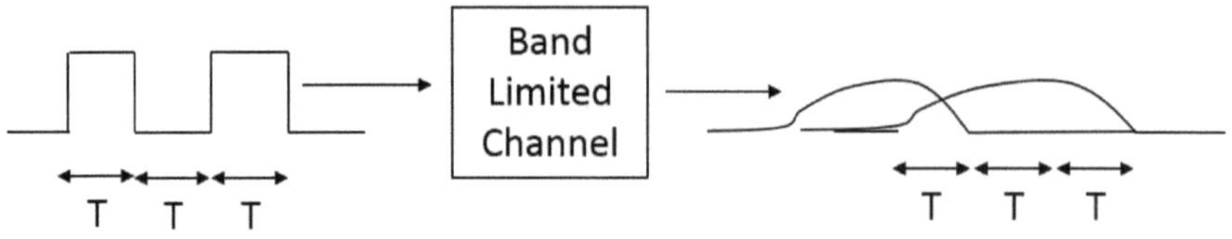

Figure 2.2.20: Pulse smearing due to band limiting

wide bandwidth spectra in the frequency domain, so it is the fast transitions in the digital signal that produce the infinite sidelobes. The corollary is that as a band limited channel removes the higher frequency components from a digital signal, the speed at which the signal can transition from one state to the other slows down. For a baseband digital signal consisting of a series of pulses, the slower transition speed at the edge of each pulse can create an effect in which the trailing energy from one pulse is "smeared" into the following pulse, as illustrated in figure 2.2.20, which creates an increased risk of error when the second pulse is detected at the receiver. This effect is called *inter-symbol interference* (ISI).

A important question is to determine the minimum bandwidth required to transmit a digital signal whilst maintaining an arbitrarily small probability of bit error. In other words, how many of the higher frequency components can be removed from a signal before the smearing effect irreparably damages the signal. This question was answered by Nyquist in 1928 [3] when he demonstrated that the minimum theoretical bandwidth required to transmit R_s symbols per second without ISI is $R_s/2$ Hertz. From figure 2.2.14 it can be seen that $R_s/2$ Hertz is half the width of the main lobe of a digital baseband signal. Nyquist showed that in order to successfully transmit a pulse stream through such a narrow channel the pulses must be shaped prior to transmission. To shape the pulses in this manner requires a square, or *brick wall,* filter in the frequency domain, as shown in figure 2.2.21(a), which shapes the time domain pulse as a sinc function shown in figure 2.2.21(b). For this example the symbol rate is four symbols per second, or the period, T, is 0.25 seconds. By restricting the bandwidth to $\frac{1}{2T}$ Hertz ($R_s/2$) the zero-crossings of the resulting sinc pulses occur every T seconds. If the pulses are transmitted at a rate of $\frac{1}{T}$ (or R_s) symbols per second, then the subsequent pulse will be sampled at exactly the same moment that the preceding pulse is passing through zero, so no ISI occurs.

In the real world it is not possible to construct a brick wall filter, since it

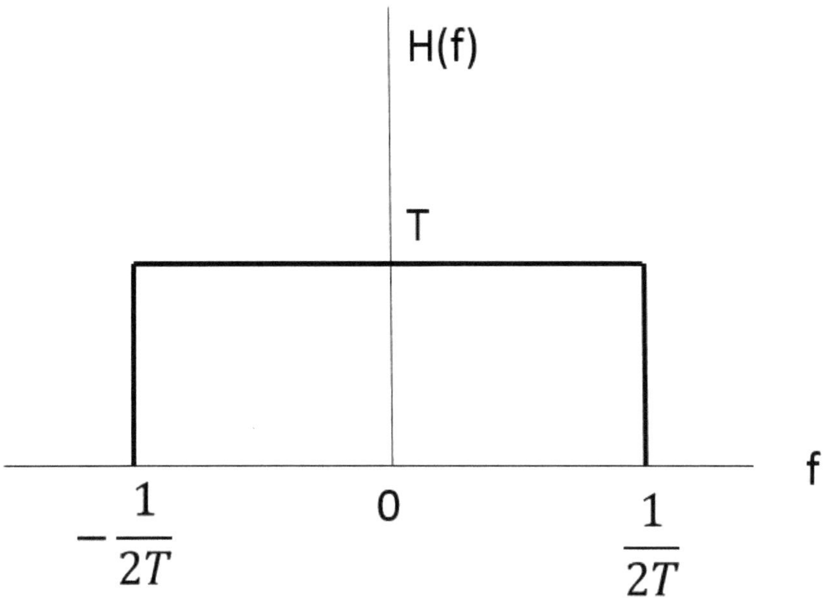

(a) Nyquist brick wall filter

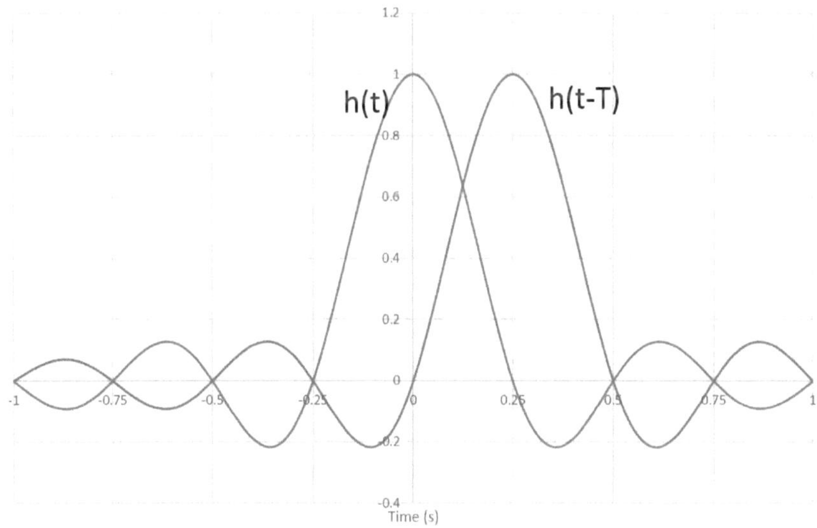

(b) Nyquist shaped pulses (T = 0.25 s)

Figure 2.2.21: Nyquist channels to avoid ISI

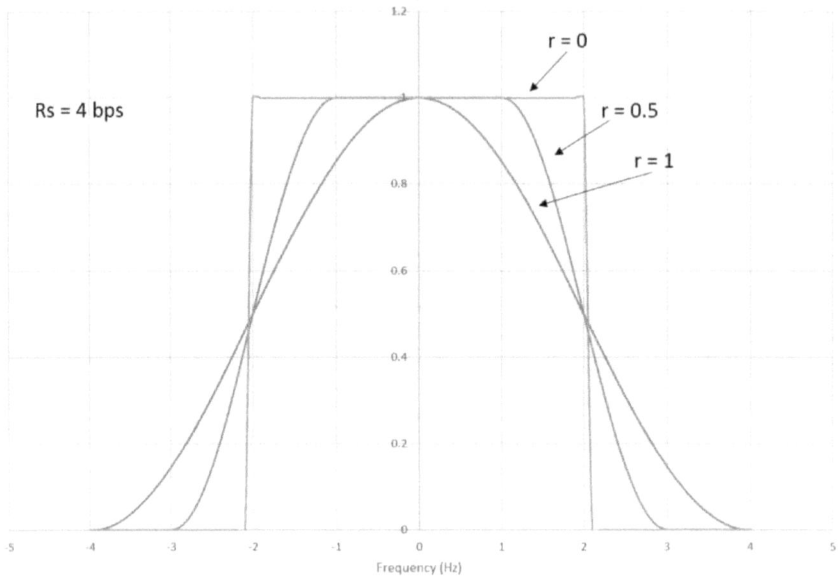

Figure 2.2.22: Raised cosine filter

would create pulses with an infinite time delay and demand exact timing to ensure that sampling occurs exactly at the zero-crossing. Typically we need to provide some excess bandwidth in order to relax these constraints, and a common alternative to the brick wall is a *raised cosine* filter. Figure 2.2.22 illustrates the shape of the raised cosine filter for various amounts of excess bandwidth as defined by the *roll-off factor*, r. For this filter, $0 \leq r \leq 1$ with $r = 0$ producing the minimum bandwidth Nyquist filter and $r = 1$ requiring double the Nyquist bandwidth. Although $r = 1$ requires more bandwidth, the timing issues are reduced because the sidelobes of the time domain sinc-like pulses produced by the filter attenuate much more quickly than the sidelobes when $r = 0$. Hence, the bandwidth required to transmit R_s symbols per second can be expressed as $W = \frac{1}{2}(1 + r) R_s$. For $r = 0$ we have the Nyquist bandwidth and for $r = 1$ we need twice the Nyquist bandwidth.

For double sideband modulation schemes, such as PSK, the minimum theoretical bandwidth to transmit R_s symbols is twice the baseband value, or R_s Hertz. However, in the real world the required bandwidth will be somewhere between R_s Hertz and $2R_s$Hertz. For example, the bit rate for a traditional VHF Combat Net Radio (CNR) is 16 kbps which, when transmitted using BPSK, requires a minimum Nyquist bandwidth of 16 kHz. But for a practical system the actual bandwidth used is somewhere between 16 kHz and 32 kHz, so it should be no surprise that the traditional standard VHF tactical channel is 25 kHz wide.

Although the Nyquist bandwidth is not practically realisable it is regularly used as a benchmark measure to compare the efficiency of one modulation scheme against another.

2.2.7 Spread Spectrum Techniques

In general, modulation schemes try to implement a bandwidth efficient method of transmitting information, since there is great competition for all parts of the radio frequency spectrum. However, there are some spread spectrum schemes which use a much wider bandwidth when compared to the symbol rate, potentially hundreds or thousands of times wider. When such schemes are properly implemented the spectrum efficiency can be close to that achieved through orthogonal modulation. Spread spectrum techniques originated with the military and were then adopted by commercial telecommunication companies for 3G mobile phone systems. In tactical systems they offer the following advantages:

1. Spreading the power produces a lower power spectral density, which makes the signal more difficult to detect.
2. Spreading the power makes the signal less susceptible to narrow band jammers.
3. Spread spectrum signals are less susceptible to multi-path interference and fading.

Many military systems use a combination of frequency spreading and time division to provide multiple, secure, narrowband channels.

Direct Sequence Spread Spectrum (DSSS)

Generating a DSSS signal is a two step process. Firstly, the (relatively) low data rate information bit stream at rate R, is combined with a high data rate *chip code* at rate R_p, using modulo 2 arithmetic. The chip code is a pseudo-random sequence that is known to both the transmitter and the receiver. The net result is a high data rate bit stream, at rate R_p, that has the low data rate information embedded inside it. The advantage of modulo 2 encoding is that at the receiver, modulo 2 adding the original chip code to the high data rate

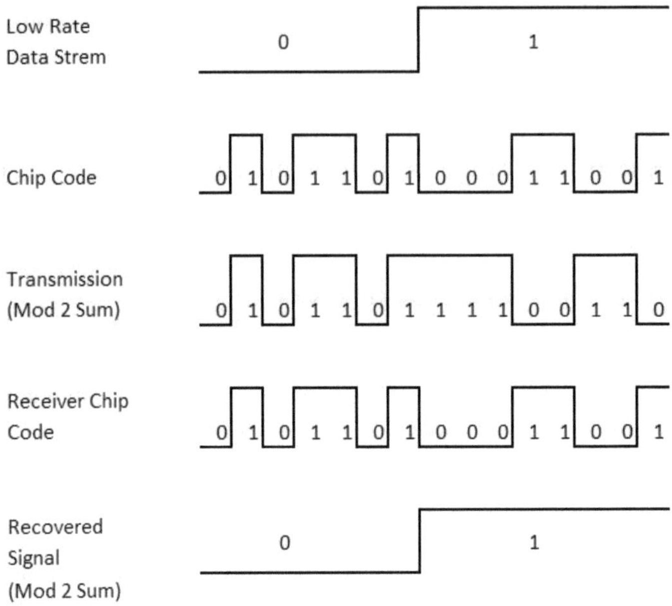

Figure 2.2.23: Modulo 2 arithmetic for DSSS

bit stream will recover the original low bit rate data stream. The process is illustrated in figure 2.2.23.

The encoded bit stream is then used to modulate a carrier frequency. We have already seen that modulating a carrier with $+1$ or -1 produces a BPSK anti-podal signal, which has a frequency spectrum similar to that in figure 2.2.15. However, for standard BPSK the bandwidth of the signal, when measured between the first nulls of the main lobe, is $2R$ Hz. For the spread spectrum signal the bandwidth is the much larger $2R_p$ Hz, yet the total transmitted power remains the same in both cases. Since the total power is represented by the area under the power spectrum graph, the net result is to dramatically lower the magnitude of the peak power components by spreading the energy across a wider bandwidth as shown in figure 2.2.24.

During transmission the DSSS signal is subject to the same AWGN that is assumed for our channel model. At first glance it may seem that the error performance would be worse, since the magnitude of the power of the signal is so much lower, but in fact that is not the case. At the receiver the demodulated signal is correlated with a replica of the spreading code, which invokes a processing gain of $G_p = \frac{R_p}{R}$. Hence, if the chip rate is one hundred times

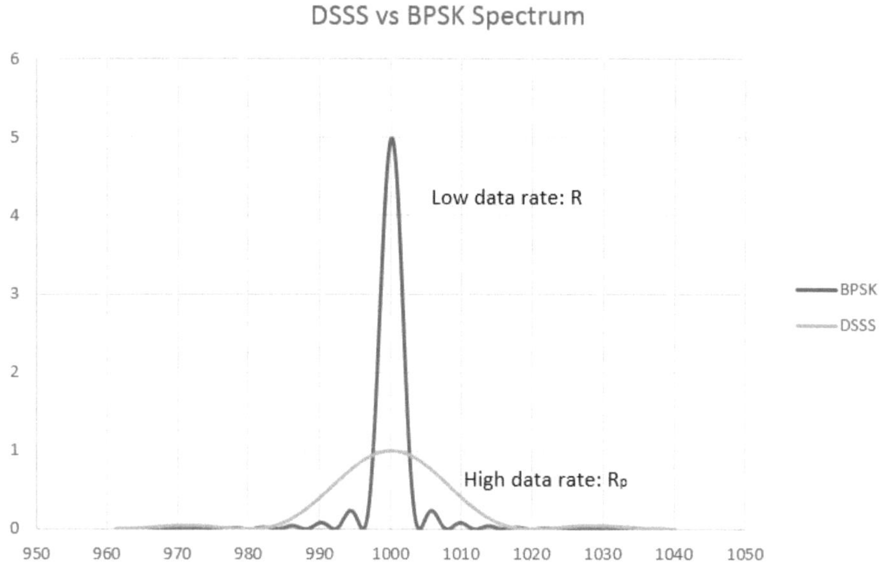

Figure 2.2.24: Power spectrum of a DSSS signal

the data rate then $G_p = 100 = 20$ dB, which is a substantial gain.

From an electronic warfare perspective DSSS offers protection against electronic attack because the jammer must decide whether to spread the jamming power across all frequencies, reducing its interference effect, or target a narrow portion of the spectrum with full power, leaving the rest of the spectrum free of interference. Intuitively, narrow band jamming will corrupt a portion of the signal energy, but most of the spread signal will be recovered through the processing gain. There is also a second effect working against the jammer. Assume that the received signal is $r(t) = s(t) + j(t)$, where $s(t)$ is the DSSS signal and $j(t)$ is the narrowband jamming signal. At the receiver $r(t)$ is modulo 2 added to the chip code which will de-spread the DSSS signal, $s(t)$. However, the narrowband jamming signal, $j(t)$, will actually have its power *spread* by the chip code, further reducing its interference effect as its power is spread into parts of the spectrum no longer occupied by the recovered narrowband $s(t)$. More detailed analysis of the electronic attack resilience of DSSS is included in chapter 8.

Frequency Hopping

Frequency hopping was the first spread spectrum technique employed in tactical radios in the late 1970s. To use frequency hopping the available spectrum is divided into a series of slots and the carrier hops between these slots at a rate known as the frequency hopping rate. If the spectrum range is W Hz, which is divided into N slots, each of bandwidth Δf, then $W = N\Delta f$. The carrier frequency at any instant is determined by the output of a frequency synthesizer, which in turn is driven by a pseudo-random code that generates one of N unique vectors into the synthesizer. The hopping rate, R_c, clocks the pseudo-random code which then determines the frequency hop rate. Note that in DSSS systems there is only a single parameter of interest, the processing gain G_p. By contrast, frequency hopping systems have two parameters of interest: the hop rate, R_C, and the hopping range, W. The bandwidth occupied during any given moment is the same as conventional M-FSK, which is typically much less than W, but averaged over time the full bandwidth W will be used. The bandwidth used in frequency hopping systems can be several GHz, which is an order of magnitude larger than the bandwidth possible in DSSS systems. Hence, the processing gain in hopping systems is usually better than DSSS systems.

Frequency hopping radio systems can be broadly classified as one of two types; *slow hopping* or *fast hopping*. A slow hopping system is one in which the hop rate is low in comparison to the data rate such that an entire symbol is modulated at one carrier frequency. A fast hopping system is one in which the hop rate is fast relative to the data rate, such that the carrier frequency changes multiple times during the transmission of one symbol. Maintaining phase lock across hops is generally not possible, so fast hopping systems require non-coherent modulation schemes. Military radios use fast hopping systems for the following reasons [1]:

1. They are resistant to enemy electronic interception because it is more difficult to follow the hopping pattern.
2. It is harder for jammers to follow the hopping pattern and disrupt the signal.
3. Fast hopping can mitigate the problems associated with fading channels through effectively employing a frequency diversity tactic.

Frequency hopping, whether fast or slow, does not affect the bit error performance of the underlying modulation scheme. More detailed analysis of the electronic attack resilience of frequency hopping is included in chapter 8.

2.2.8 Conclusion

This section has focussed on the physical layer of a tactical radio circuit which is governed by the nature of the channel. Radio spectrum allocation affects the available bandwidth, which is intimately tied to the sustainable bit rate, and signal fading in radio channels must also be addressed. At the physical layer the modulation scheme is selected that optimises the competing priorities of available bandwidth, available power to combat the bit error rate, and the required bit rate to service the information needs of the end users. Electronic resilience is also implemented at the physical layer through spread spectrum techniques such as frequency hopping and DSSS.

Although peer-to-peer data circuits exist at the transport layer, they are established by a series of point-to-point hops across the network. The physical layer addresses some of the issues involved in achieving an individual hop. In the next section we will examine the multiple access control layer, which allows multiple nodes to share the physical layer resources across each hop. Then section 2.4 will consider the Data Link Layer (DLL) which introduces channel coding on the bit stream prior to modulation in order to improve the resilience to noise of the transmission across the individual point-to-point hop.

2.3 Multiple Access Control Layer in Tactical Networks

2.3.1 Carrier Sense Multiple Access (CSMA)

Tactical communications is built around mobile, ad-hoc networks (MANET) with an underlying radio implementation at nodes separated by sometimes tens of kilometres. Radio nets are *all informed*, in that when one person transmits all of the other nodes on the network will receive the transmission. This has been a traditional advantage for voice radio nets because it provides situational awareness to all net participants. Clearly it is important that only one person transmits at any given time in order to avoid collisions on the net which will distort the transmission. For voice nets there is a radio operator at each node who listens to the net traffic, and when the net is free transmits the message. This is a simple implementation of the Carrier Sense Multiple Access (CSMA) protocol in which a node *senses* whether or not the carrier signal is being transmitted and then, if there is no activity, transmits their message.

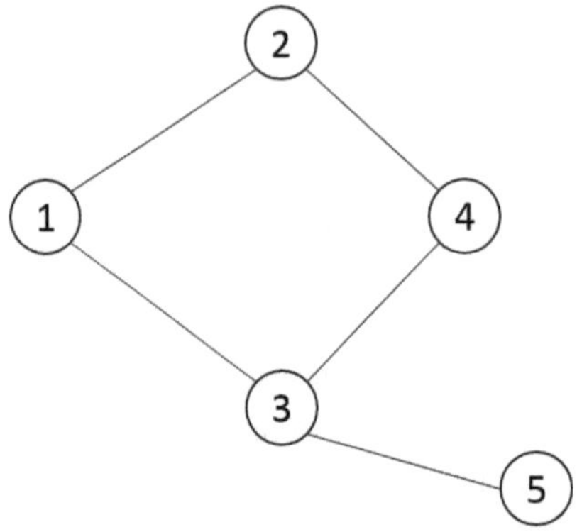

Figure 2.3.1: Node reachability graph

On commercial data networks the Ethernet protocol uses a CSMA approach to share the local area network bandwidth. However, extending CSMA to tactical networks is not practical for MANET. The drawback with CSMA schemes is that collisions do occur on the network, which usually requires a re-transmission, thus reducing the information throughput. Consider the node diagram in figure 2.3.1, in which the lines between nodes indicate which nodes are able to communicate with which other nodes. If node five is transmitting to node three, and simultaneously node one wishes to transmit to node three, then node one *senses* for carrier activity. But for this network node one will not be able to hear the activity of node five, and therefore will commence transmission. At this point node three is receiving from both node five and node one and a collision has occurred. Both nodes will need to re-transmit their data, although there is a risk of repeating another collision. In general, CSMA protocols can only support around 50% of the available bandwidth [4].

2.3.2 Frequency Division Multiple Access (FDMA)

In FDMA a portion of the electromagnetic spectrum is allocated to each transmitter, which can then transmit in the allocated channel as required. The advantage of FDMA is that no central control station is required and close to 100% of the available channel capacity can be used. There is no danger of one transmission disrupting another, and there is no delay introduced by the channel. The disadvantage is that allocation of the spectrum resource is fairly inflexible and it is difficult for one node to receive data from more than one transmitter, since multiple receivers are required.

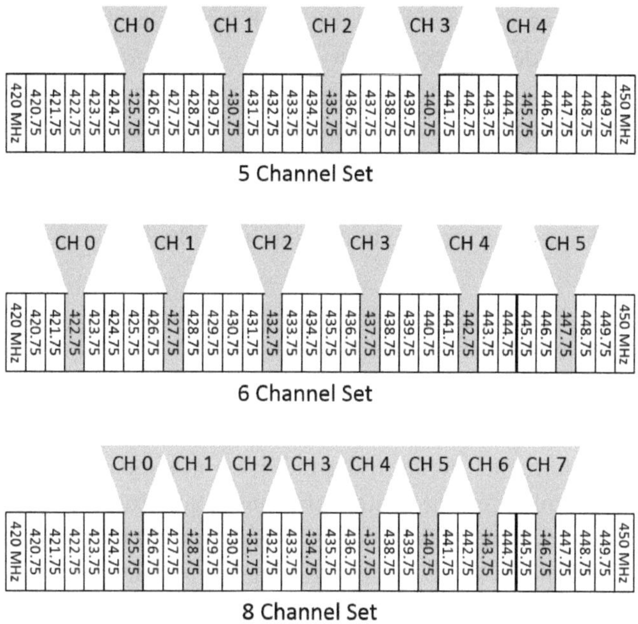

Figure 2.3.2: EPLRS FDMA structure

FDMA is often used to separate multiple user groups by establishing narrow band nets at separate points in the frequency spectrum. However, it is best suited for point-to-point radio links and is often used for satellite services. Frequency hopping is a form of FDMA in which orthogonal hopping patterns can be used to separate radio nets by frequency. Figure 2.3.2 shows the FDMA scheme for the EPLRS radio system, where five, six or eight frequency channels are used to separate different formation networks.

2.3.3 Code Division Multiple Access (CDMA)

As a multiple access technique CDMA has found more popularity in commercial applications, rather than military applications. It works well for establishing multiple point-to-point links to a central node, such as a fixed base station, and was a backbone technology for 3G mobile services. The technology behind CDMA is the same as for Direct Sequence Spread Spectrum, in that a low data rate digital voice transmission is used to modulate a high data rate chip code and create a spread spectrum signal with low peak power spectrum. Multiple users are accommodated by giving each mobile user a carefully selected unique chipping code. It is possible to derive *orthogonal* chip codes such that at the receiver no user interferes with the signal of another user. Each node uses the full bandwidth for each transmission, producing a low power signal over the whole bandwidth. When multiple users are transmitting the signals are additive, but the orthogonal chip codes provide sufficient processing gain for the base station receiver to extract them individually.

2.3.4 Time Division Multiple Access (TDMA)

In a TDMA system each node in the radio net is allocated a fixed time slot in which to transmit. Central control of timing synchronisation is required in order to avoid drift in the transmissions, or at least regular transmissions from each station. During its allocated time slot the transmitter uses the entire available bandwidth, but it must then remain silent for the remaining time. The advantages of TDMA are that it is easy for a station to monitor all timeslots and remain all-informed; there is no data loss from one station transmitting over another and there is a fixed upper bound on delay, since a station need only wait until his next time window cycles around again. It also supports spread spectrum techniques because the whole bandwidth is used. The disadvantages are that timing synchronisation between stations must be maintained; guard intervals are required for multiple node nets and allocation of capacity can be inflexible [4]. TDMA has been used for many years in tactical networks such as LINK-16 and the Enhanced Position Location Reporting System (EPLRS).

2.3.5 Demand Assignment Multiple Access (DAMA)

DAMA is a variant of TDMA in which statistical multiplexing of nodes is used. On a DAMA network there are usually more nodes than time slots available. When a node needs to transmit it requests access to a time slot via a control channel. The net control station allocates a time slot to the node, based on a priority of service algorithm. Once the slot is allocated, the node transmits its data in that slot and then surrenders the slot when it has completed its transmission.

2.4 The Data Link Layer

The physical and MAC layers of the tactical network protocol stack addresses some specific transmission issues that arise simply due to the nature of radio communications between two points. The Data Link Layer (DLL) is the first layer in the protocol stack that focusses on preserving the *information* in the bit stream, rather than the bits themselves. Channel coding introduces additional bits into the signal in order to detect, and ideally correct, one or more bit errors at the receiver.

In tactical data networks channel coding is complex because each of the protocol layers acts independently. The DLL can provide channel coding on a hop-by-hop basis, whilst the transport layer can also provide end-to-end channel coding. Exactly where in the protocol stack channel coding is best implemented is a current research topic. Some researchers promote network coding to make sure that each hop adapts dynamically to the channel conditions, ensuring that all the network hops are reliable and hence reduce the burden on the transport layer [1]. Others advocate cross-layer signalling to help the higher layers implement information theory strategies, while others propose proxy protocols that adapt to changes in the wireless network while COTS applications are used seamlessly by the end-user.

2.4.1 Channel Coding for Error Correction

Channel coding is the process of adding additional bits to a message stream, often called redundancy, in order to improve the bit error rate of the message. All tactical data radios employ some form of error correcting code in order to compensate for the harsh electromagnetic environment. It is doubtful whether meaningful tactical data rates could be achieved without the power provided through these sophisticated techniques.

The extra bits help to detect errors in transmission and potentially also correct the corrupted bit or bits. If the coding only provides error detection, then a repeat request (ARQ) strategy is used each time an error is detected. However, most schemes also include the capability to correct the corrupted bit or bits at the receiver, without requiring a re-transmission, and this approach is called Forward Error Correction (FEC). In general, if a k bit message block is coded into an n bit codeword then the scheme is called an (n, k) code. The coding rate is $\frac{k}{n}$ which indicates the proportion of information transmitted in each codeword. Hence, if a rate $\frac{1}{2}$ code is used then 50% of the bits transmitted are overhead for correcting purposes. If rate $\frac{3}{4}$ is used then 25% of the transmitted bits are error correcting overhead.

Orthogonal Coding

Wireless data communications is an ongoing trade-off between power, bandwidth and error performance, and we have already seen that we can improve the error performance of a modulation scheme if we transmit orthogonal symbols. For example, if the probability of a symbol error in a channel is one in ten symbols, then for binary FSK, which transmits one bit per symbol, $P_s = P_b = 0.1$. However, an orthogonal M-ary modulation scheme, such as 8-FSK, can improve that error performance at the expense of bandwidth. In an 8-FSK system, three bits are transmitted per symbol, so if one symbol is corrupted, causing perhaps one of the three bits to be incorrectly decoded, then $P_s = 0.1$, but $P_b = 1/30 = 0.03$. Error coding, or channel coding, can have a similar effect if we transform the message bits (length k) into codewords (length n) through a structured process.

From a geometric perspective we can consider the message block to be a k-tuple, or a k-dimensional vector and each codeword to be an n-tuple, forming a set of 2^k n-dimensional vectors. Of course, the entire set of 2^n possible n-dimensional vectors forms a vector space, and to be considered a linear block code the selected codeword subset must have the properties of a vector *subspace*. Principally, the 2^k codeword subset must contain the all-zero vector and the set must be *closed.* This means that the summation of any two codewords in the set will produce another codeword from the set[3]. If the codeword subset is a *subspace* then it turns out that each valid codeword will be orthogonal to every non-codeword n-dimensional vector, which is where the codewords draw their error correcting strength. There are broadly two classes of codes: linear block codes and convolutional codes. Both types are used in tactical radios, as well as a concatenation of block and convolutional coding to provide particular resilience against burst noise. Figure 2.4.1 illustrates how error coding and modulation are related in the communications process.

[3]Summation of binary numbers uses modulo 2 addition. That is, $0 \oplus 0 = 0, 0 \oplus 1 = 1, 1 \oplus 0 = 1, 1 \oplus 1 = 0$.

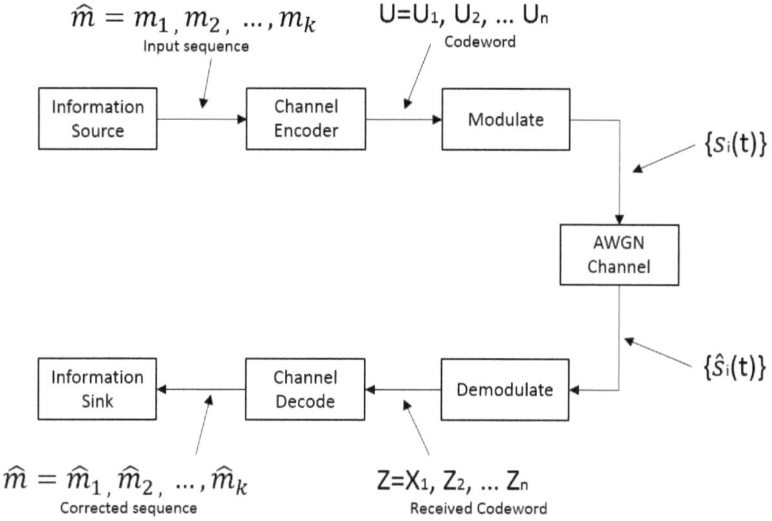

Figure 2.4.1: Error coding and modulation

2.4.2 Linear Block Codes

Block codes map a fixed length message block into a fixed length codeword. The coding is *memoryless* because the codes are not influenced by the previous message bits, but only by the current k-tuple of message bits.

Generation

Consider a (6,3) linear block code that maps 3-tuples of message data into 6-tuple codewords using the mapping listed in table 2.2. It is straightforward to confirm that the codeword set is a subspace of the 2^6 possible binary 6-tuples. It contains the zero vector and adding any codewords together produces another codeword in the set.

There are, of course, a large number of possibilities that one could choose for the mapping. The codewords form a three dimensional subspace within the six dimensional space, but there are twenty possible three dimensional subspaces that could be selected. An obvious question is whether one subspace offers better error recovery power than another. This has been a fruitful research area for many years which has demonstrated that certain mappings do have better properties than others.

Data 3-tuple	Codeword Set
000	000000
100	110100
010	011010
110	101110
001	101001
101	011101
011	110011
111	000111

Table 2.2: (6,3) linear block code

For this simple code a look-up table is adequate to perform the mapping, but many codes involve much longer block lengths, such as a (127, 92) code for which there are 2^{92} (or 5×10^{27}) code vectors. For such systems we need a generation technique that takes the 92-tuple input vector and generates the 127-tuple code vector on-the-fly. The key is to recognise that for an (n,k) code the subspace of 2^k code vectors is spanned by just k vectors. That is, we can produce any code vector $\mathbf{u}$ as a linear combination of the k basis code vectors $\mathbf{v}_i$, or $\mathbf{u} = m_1\mathbf{v}_1 + m_2\mathbf{v}_2 + \cdots + m_k\mathbf{v}_k$. Therefore, if we identify the k basis vectors, we can write down a generator matrix in the form

$$\mathbf{G} = \begin{bmatrix} \mathbf{v}_1 \\ \mathbf{v}_2 \\ \vdots \\ \mathbf{v}_k \end{bmatrix}$$

and then, for a message sequence $\mathbf{m} = m_1, m_2, \ldots, m_k$ the corresponding code vector is given by $\mathbf{u} = \mathbf{mG}$.

Example. The basis vectors for the mapping in table 2.2 are

$$\mathbf{G} = \begin{bmatrix} 1 & 1 & 0 & 1 & 0 & 0 \\ 0 & 1 & 1 & 0 & 1 & 0 \\ 1 & 0 & 1 & 0 & 0 & 1 \end{bmatrix}$$

and hence the codeword for the input block [110] is

$$\begin{aligned}\mathbf{u} &= \mathbf{mG} \\ &= \begin{bmatrix} 1 & 1 & 0 \end{bmatrix} \begin{bmatrix} 1 & 1 & 0 & 1 & 0 & 0 \\ 0 & 1 & 1 & 0 & 1 & 0 \\ 1 & 0 & 1 & 0 & 0 & 1 \end{bmatrix} \\ &= \begin{bmatrix} 1 & 0 & 1 & 1 & 1 & 0 \end{bmatrix}\end{aligned}$$

Error Detection

To decode a received codeword and detect an error we must first create the parity check matrix $\mathbf{H}$. This $(n-k) \times n$ matrix spans the null space of $\mathbf{G}$ which means that the rows of $\mathbf{H}$ are each orthogonal to every row in $\mathbf{G}$. Deriving $\mathbf{H}$ from $\mathbf{G}$ is a straightforward mathematical process [2].

Let the received code vector be $\mathbf{z} = \mathbf{u} + \mathbf{e}$ where $\mathbf{e}$ is an error n-tuple that corrupts $\mathbf{u}$ during transmission. We calculate the *syndrome* of $\mathbf{z}$ as

$$\begin{aligned}\mathrm{s} &= \mathbf{zH}^T \\ &= (\mathbf{u}+\mathbf{e})\,\mathbf{H}^T \\ &= \mathbf{uH}^T + \mathbf{eH}^T. \end{aligned} \tag{2.4.1}$$

Now $\mathbf{uH}^T = \mathbf{0}$ because the codeword $\mathbf{u}$ is a linear combination of the rows of $\mathbf{G}$ and the rows of $\mathbf{G}$ are all orthogonal to the columns of $\mathbf{H}^T$ (ie, the rows of $\mathbf{H}$). Therefore, if the syndrome $\mathrm{s} = \mathbf{0}$ then $\mathbf{e} = \mathbf{0}$ and $\mathbf{z}$ is a valid codeword and should be decoded. If $\mathbf{S} \neq \mathbf{0}$ then $\mathbf{e}$ is non-zero and an error has been detected.

Example. Let us assume that the codeword $\mathbf{u} = \begin{bmatrix} 1 & 0 & 1 & 1 & 1 & 0 \end{bmatrix}$ has been transmitted, but it has been corrupted by the error vector $\mathbf{e} = \begin{bmatrix} 1 & 0 & 0 & 0 & 0 & 0 \end{bmatrix}$ producing the received codeword

$$\mathbf{z} = \begin{bmatrix} 0 & 0 & 1 & 1 & 1 & 0 \end{bmatrix}.$$

The $(n-k) \times n$ parity matrix is

$$\mathbf{H} = \begin{bmatrix} 1 & 1 & 1 & 0 & 0 & 0 \\ 1 & 0 & 0 & 1 & 0 & 1 \\ 0 & 1 & 0 & 1 & 1 & 0 \end{bmatrix}$$

and the syndrome of $\mathbf{z}$ is

$$
\begin{aligned}
\mathrm{s} &= \begin{bmatrix} 0 & 0 & 1 & 1 & 1 & 0 \end{bmatrix} \begin{bmatrix} 1 & 1 & 0 \\ 1 & 0 & 1 \\ 1 & 0 & 0 \\ 0 & 1 & 1 \\ 0 & 0 & 1 \\ 0 & 1 & 0 \end{bmatrix} \\
&= \begin{bmatrix} 1 & 1 & 0 \end{bmatrix}.
\end{aligned}
$$

Therefore, since s is non-zero an error has been detected.

Error Correction

Calculating a non-zero syndrome indicates that an error has occurred, and the next step is to correct the error. Equation 2.4.1 indicates that the syndrome calculated for a corrupted codeword, $\mathbf{z}$, is identical to the syndrome produced by the error vector, $\mathbf{e}$. Therefore, it is possible to use a lookup table, such as table 2.3, to match the syndrome results to the error vector that caused the corruption. Once the error vector is identified then the correct codeword can be recovered by simply subtracting the error from $\mathbf{z}$. Of course, with modulo 2 arithmetic subtraction and addition are the same process, so in practice we simply add the error vector to $\mathbf{z}$ in order to recover $\mathbf{u}$.

Note that there are 2^{n-k} possible syndrome results, which for this example means that eight syndrome patterns are available. One of them is the zero vector and the next six are the single-bit errors, which is a total of seven patterns. Therefore, there is still one more syndrome pattern available which can be used to test for a two-bit corruption. The choice of which two-bit corruption to correct is arbitrary.

Example. The calculated syndrome is $\mathbf{S} = \begin{bmatrix} 1 & 1 & 0 \end{bmatrix}$ which indicates, from table 2.3, that the error vector is $\mathbf{e} = \begin{bmatrix} 1 & 0 & 0 & 0 & 0 & 0 \end{bmatrix}$. Therefore the correct codeword is recovered by modulo 2 adding $\mathbf{e}$ to $\mathbf{z}$.

$$
\begin{aligned}
\mathbf{u} &= \mathbf{z} \oplus \mathbf{e} \\
&= \begin{bmatrix} 0 & 0 & 1 & 1 & 1 & 0 \end{bmatrix} \oplus \begin{bmatrix} 1 & 0 & 0 & 0 & 0 & 0 \end{bmatrix} \\
&= \begin{bmatrix} 1 & 0 & 1 & 1 & 1 & 0 \end{bmatrix}
\end{aligned}
$$

Error Pattern	Syndrome
0 0 0 0 0 0	0 0 0
0 0 0 0 0 1	0 1 0
0 0 0 0 1 0	0 0 1
0 0 0 1 0 0	0 1 1
0 0 1 0 0 0	1 0 0
0 1 0 0 0 0	1 0 1
1 0 0 0 0 0	1 1 0
0 1 0 0 0 1	1 1 1

Table 2.3: Syndrome lookup table

2.4.3 Coding Strength and Coding Gain

Choosing the "best" set of codewords has been an active research area. The strength of a code to correct errors is governed by how *different* each codeword is from all of the other codewords in the set. The more different they are, the less likely a corrupted codeword will be wrongly assigned to a valid codeword. This concept of difference is characterised as the *distance* between each pair of codewords. The error performance will be governed by the smallest distance between a pair of codewords in the set, which is called the *minimum distance,* d_{min} of the set.

The *Hamming distance* between two codewords $\mathbf{u}$ and $\mathbf{v}$, denoted $d(\mathbf{u}, \mathbf{v})$, is defined as the number of elements in which they differ. For example, if $\mathbf{u} = \begin{bmatrix} 1 & 0 & 0 & 1 & 1 \end{bmatrix}$ and $\mathbf{v} = \begin{bmatrix} 1 & 1 & 0 & 1 & 0 \end{bmatrix}$ then the Hamming distance is $d(\mathbf{u}, \mathbf{v}) = 2$. The number of correctable codewords, t, is given by $t = \left\lfloor \frac{d_{min}-1}{2} \right\rfloor$ where $\lfloor x \rfloor$ means the largest integer not to exceed x. The number of guaranteed detectable codewords is $d_{min} - 1$.

Figure 2.4.2 shows the probability of bit error, P_B, versus $\frac{E_b}{N_0}$ for coherent binary PSK modulation in combination with examples of various (n, k) codes over a Gaussian channel [6]. The (1,1) curve is uncoded performance, while the (24, 12) and the (127, 92) curves illustrate coded PSK performance using block codes with $(n - k) = 12$ parity bits and 35 parity bits, respectively. The curves illustrate an important feature of channel coding. Coding gain is not realised unless sufficient power is available to the system. In this example we require $\frac{E_b}{N_0} \approx 5.5$dB before there is any benefit derived from the block codes. In fact, for power less than this figure the block codes degrade the

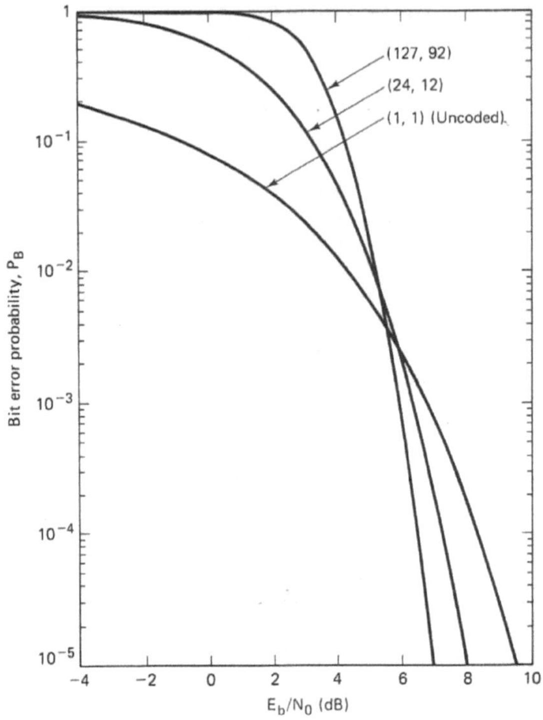

Figure 2.4.2: Probability of bit error for PSK with various (n,k) codes [6]

error performance of the BPSK circuit because they are simply overhead. The degradation occurs because we have the same finite power now being spread over n bits instead of k bits, so we have reduced energy per bit when compared to the uncoded case. However, once the threshold is exceeded the performance improvement through error detection and correction more than compensates for the reduced energy per bit.

Coding Gain is the reduction in the required $\frac{E_b}{N_0}$ to achieve a specified error performance of an error correcting coded system over an uncoded one with the same modulation. For example, in figure 2.4.2, for $P_B = 10^{-5}$ the (24,12) code has a coding gain of around 1.5 dB.

2.4.4 Common Block Codes

Hamming Codes

Hamming codes have the structure $(n, k) = (2^m - 1, 2^m - 1 - m)$, where $m = 2, 3 \ldots$. These codes have a minimum distance of three and are capable of correcting all single errors or detecting combinations of two or fewer errors within a block.

Extended Golay Codes

The Extended Golay code is a $(24, 12)$ code formed by adding a parity bit to the $(23, 12)$ Golay code. This added parity bit increases the minimum distance from seven to eight and produces a rate $\frac{1}{2}$ code that is easier to implement than the rate $\frac{12}{23}$ of the original Golay code [6]. Extended Golay codes are more powerful than Hamming codes, but require a more complex decoder, produce a lower code rate and hence require additional bandwidth to maintain the underlying data rate.

BCH Codes

Bose-Chadhuri-Hocquenghem (BCH) codes are a generalisation of Hamming codes that allow multiple error correcting. They are powerful cyclic codes [6] with a large selection of block lengths, code rates, alphabet sizes and error correcting capability. Alphabet variation means that non-binary alphabets can be encoded using BCH codes. For example, an (n, k) code maps k message symbols onto n code symbols. When a binary alphabet is used each of the k symbols is a single bit, but for a non-binary alphabet each of the k symbols represents m bits, where m is the alphabet size.

BCH codes generally use long block lengths because at block lengths of a few hundred symbols the BCH codes outperform all other block codes with the same block length and code rate [6].

Reed-Solomon Codes [5]

Reed-Solomon codes are a particular sub-class of BCH codes that have been shown to be particularly effective for burst error recovery, or channels that have memory. Tactical radio channels fall into this category and Reed-Solomon

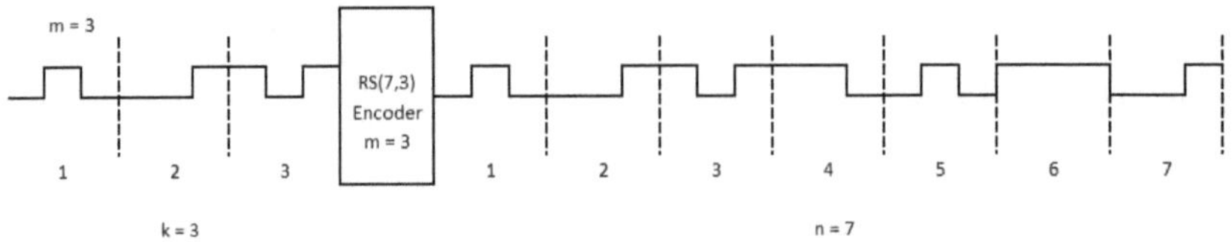

Figure 2.4.3: RS(7,3) code with $m = 3$ bits per symbol

codes are widely deployed in current tactical radio systems. Reed-Solomon codes are non-binary cyclic codes with each symbol made up of m-bit sequences, where m is any positive integer having a value greater than 2. Reed-Solomon (n, k) codes on m-bit symbols exist for all n and k for which $0 < k < n < 2m + 2$ where k is the number of data symbols being encoded, and n is the total number of code symbols in the encoded block. For the most conventional Reed-Solomon (n, k) code, $(n, k) = (2m - 1, 2m - 1 - 2t)$ where t is the symbol-error correcting capability of the code, and $n - k = 2t$ is the number of parity symbols. Reed-Solomon codes achieve the largest possible code minimum distance for any linear code with the same encoder input and output block lengths. Figure 2.4.3 illustrates a RS(7,3) code with $m = 3$ bits per symbol.

The code minimum distance is given by $d_{min} = n - k + 1$ The code is capable of correcting any combination of t or fewer errors, where t can be expressed as $t = \left\lfloor \frac{d_{min}-1}{2} \right\rfloor = \left\lfloor \frac{n-k}{2} \right\rfloor$. Thus, for the case of Reed-Solomon codes, correcting t symbol errors requires no more than $2t$ parity symbols. An advantage of non-binary codes such as a Reed-Solomon code can be seen by the following comparison. Consider a binary $(n, k) = (7, 3)$ code. The entire n-tuple space contains $2^n = 2^7 = 128$,n-tuples, of which $2^k = 2^3 = 8$ (or 1/16 of the n-tuples) are codewords. Next, consider a non-binary $(n, k) = (7, 3)$ code where each symbol is composed of $m = 3$ bits. The n-tuple space amounts to $2^{nm} = 2^{21} = 2,097,152$ n-tuples, of which $2^{km} = 2^9 = 512$ (or 1/4096 of the n-tuples) are codewords. When dealing with non-binary symbols, each made up of m bits, only a small fraction (i.e., 2^{km} of the large number 2^{nm}) of possible n-tuples are codewords. This fraction decreases with increasing values of m. The important point here is that when a small fraction of the n-tuple space is used for codewords, a large d_{min} can be created.

To show why Reed-Solomon codes are effective against burst transmission, consider an $(n, k) = (255, 247)$ Reed-Solomon code, where each symbol is

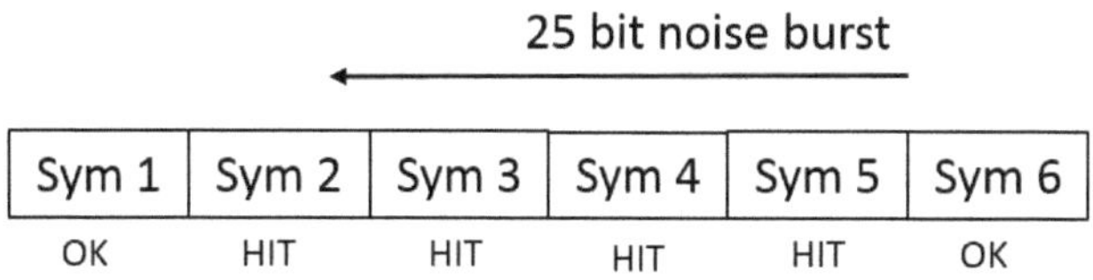

Figure 2.4.4: Data block disturbed by 25 bit noise burst

made up of $m = 8$ bits. Since $n - k = 8$, this code can correct any four symbol errors in a block of 255. Imagine the presence of a noise burst, lasting for 25-bit durations and disturbing one block of data during transmission, as illustrated in figure 2.4.4.

In this example, notice that a burst of noise that lasts for a duration of 25 contiguous bits must disturb exactly four symbols. The Reed-Solomon decoder for the $(255, 247)$ code will correct any four-symbol errors without regard to the type of damage suffered by the symbol. In other words, when a decoder corrects a byte, it replaces the incorrect byte with the correct one, whether the error was caused by one bit being corrupted or all eight bits being corrupted. Thus if a symbol is wrong, it might as well be wrong in all of its bit positions. This gives an Reed-Solomon code a tremendous burst-noise advantage over binary codes, even allowing for the interleaving of binary codes.

2.4.5 Convolutional Codes

The second major error coding scheme used in tactical radios is convolutional coding. Whereas block codes are characterised by two numbers, (n, k), convolutional encoding is described by three numbers, n, k, and K, where $\frac{k}{n}$ has the same significance as the code rate, but n no longer represents a block length. The third number, K, is called the constraint length and it represents the number of k-tuple stages in the shift register used to generate the code. The important difference in convolutional coding is that the encoder has memory. That is, the output depends not just on the current input, but also on the last $K - 1$ inputs. Typically, n and k are small and K is varied to control the redundancy.

An example of a convolutional encoder is shown in figure 2.4.5, which shows a $(2, 1)$ convolutional encoder with constraint $K = 3$. For each single input bit $(k = 1)$ two output bits are sampled from the modulo 2 adders $(n = 2)$, producing a *branch word* [6]. As successive bits enter the encoder on the left,

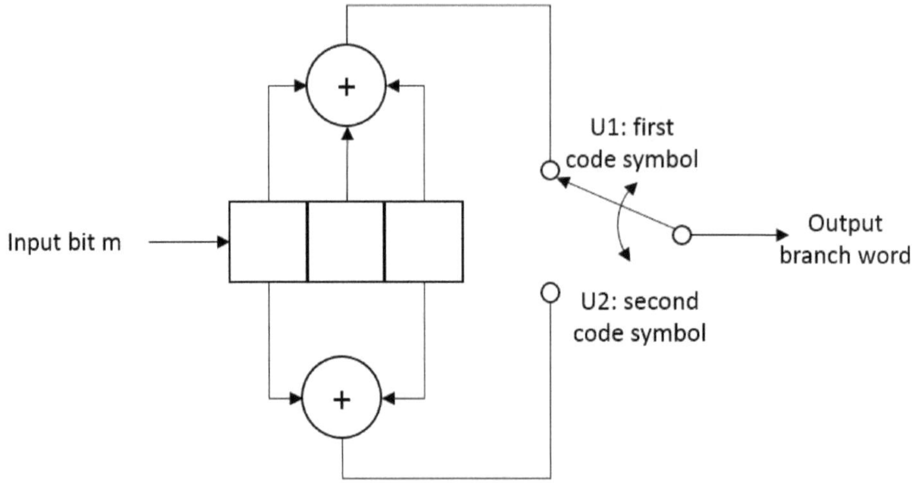

Figure 2.4.5: Convolutional encoder (rate $\frac{1}{2}$, $K = 3$)

the first bit is shifted across to the right so that it continues to contribute to the output during the second and third clock cycles. After $K = 3$ clock cycles the first bit no longer contributes to the output and it has dropped out of the encoder's memory. Each input bit produces two output bits, and hence the code rate is $\frac{1}{2}$, but n does not represent a block length.

Example. Consider a message vector $\mathbf{m} = \begin{matrix} 1 & 0 & 1 \end{matrix}$ that is convolutionally encoded using the encoder in figure 2.4.5. At t_1,t_2 and t_3 the message is clocked into the encoder and then at t_4 and t_5 two zeros are clocked into the encoder in order to flush the last message bit all the way through the shift registers. As each bit is clocked in, two output bits are read from the modulo 2 adders. Table 2.4 shows the output from the encoder as the message bits are clocked through the shift registers. The complete output sequence is $\begin{matrix} 1 & 1 & 1 & 0 & 0 & 0 & 1 & 0 & 1 & 1 \end{matrix}$. Notice that the effective code rate is $\frac{3}{10}$, which is less than the expected $\frac{1}{2}$. This is due to the need to flush the registers with two zeros, which are redundant. For this small example the effect is significant, but for longer code streams the extra two zeros are negligible.

Polynomial Representation

In the literature, convolutional encoders are frequently described using polynomials to represent the connections to each modulo 2 adder. Polynomial

Time	Shift Register	Output
t_1	1 0 0	1 1
t_2	0 1 0	1 0
t_3	1 0 1	0 0
t_4	0 1 0	1 0
t_5	0 0 1	1 1

Table 2.4: Example Encoding of $\mathbf{m} = \begin{array}{ccc} 1 & 0 & 1 \end{array}$

notation is also used in MIL-STD-188-220, which specifies the acceptable error coding schemes for data over combat net radio. Each polynomial is of degree $K-1$ or less and there are n polynomials to characterise the encoder, one for each adder.The coefficient of each term in the polynomial is either a one or a zero, depending on whether that shift register element is connected to the adder. For the example in figure 2.4.5 the upper and lower adders are represented by the following polynomials

$$\begin{aligned} \mathbf{g}_1(X) &= 1+X+X^2 \\ \mathbf{g}_2(X) &= 1+X^2 \end{aligned}$$

where the lowest term in the polynomial corresponds to the input register. This representation can be used to exploit the algebraic structure in a convolutional encoder. The output sequence is $\mathbf{u}(X) = \mathbf{m}(X)\mathbf{g}_1(X)$ interlaced with $\mathbf{m}(X)\mathbf{g}_2(X)$.

Example. First express the message vector $\mathbf{m} = \left[\begin{array}{ccc} 1 & 0 & 1 \end{array}\right]$ as a polynomial: $\mathbf{m}(X) = 1 + X^2$.

$$\begin{aligned} \mathbf{m}(X)\mathbf{g}_1(X) &= (1+X^2)(1+X+X^2) = 1+X+X^3+X^4 \\ \mathbf{m}(X)\mathbf{g}_2(X) &= (1+X^2)(1+X^2) = 1+X^4 \end{aligned}$$

hence

$$\begin{aligned} \mathbf{m}(X)\mathbf{g}_1(X) &= 1+\ \ X+0X^{2+}\ X^3+X^4 \\ \mathbf{m}(X)\mathbf{g}_2(X) &= 1+0X+0X^2+0X^3+X^4 \\ \mathbf{u}(X) &= (1,1)+(1,0)X+(0,0)X^2+(1,0)X^3+(1,1)X^4 \\ &= \begin{array}{ccccc} 1\ 1 & 1\ 0 & 0\ 0 & 1\ 0 & 1\ 1 \end{array} \end{aligned}$$

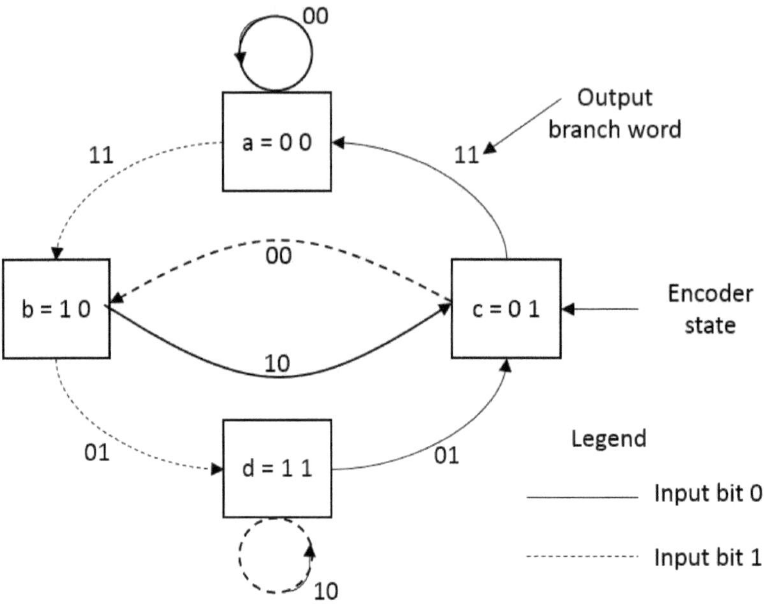

Figure 2.4.6: Encoder state diagram (rate $\frac{1}{2}$, $K = 3$)

State Representation

The state of a convolutional encoder is defined as the contents of the right-most $K - 1$ stages. As a message is encoded the state of the encoder changes, but in a structured way which is called a *Markov chain*. A characteristic of Markov chains is that the output at time t is governed only by the input at time t and knowledge of the current state. The history of how the chain reached the current state has no influence on the output at time t. A state diagram for the encoder in figure 2.4.5 is shown in figure 2.4.6. The four possible states are named a, b, c, and d and the arrows indicate the possible state transitions that can occur. Note that not every transition is possible. For example, if the encoder is in state $c = 01$ then it cannot enter state $d = 11$ in a single transition. The arrows indicate the type of input bit (0 or 1) and the numbers alongside the arrows indicate the resulting output pair.

2.4.6 Decoding Convolutional Codes

Decoding block codes is achieved by testing the syndrome of a received codeword and using that syndrome to identify which bit in the block has been corrupted and then repairing that bit. By contrast, convolutional codes assess a large length of the received sequence and try to deduce the most likely transmitted sequence to have produced that received sequence. This approach introduces some latency, because we must sometimes delay the decision on the first few bits of a sequence until we have received a substantial length of the entire sequence.

This approach is called maximum likelihood decoding because it seeks to determine the sequence that has the maximum likelihood of being the one that generated the received sequence. The most common algorithm is called the Viterbi decoding algorithm, named after Andrew Viterbi who proposed it in 1967 [7]. The Viterbi algorithm uses the Markov structure, which restricts certain state transitions, in order to assess which sequence is most likely, since certain sequences cannot occur. However, at each time t_i, if two sequences enter the same encoder state then their future trajectories will be identical, because of the Markov structure of the encoder. Therefore, the Viterbi algorithm measures the distance between the received sequence and the two candidate sequences up to that instant and deletes the sequence with the larger distance. Deleting sequences on-the-fly reduces the computations required considerably and makes this maximum likelihood algorithm tractable.

Trellis Representation

A trellis diagram can be used to illustrate the Viterbi algorithm. The trellis diagram uses the fact that each encoder state has no history, in order to simplify the comparison between the received sequence and all of the possible generator sequences. An example of a trellis diagram is shown in figure 2.4.7 for the encoder in figure 2.4.5. The state of the encoder is listed on the left, with two possible outputs from each state, depending on whether the input is a zero or a one. An input zero is indicated by a solid line and an input one is indicated by a dashed line. For example, if the encoder is in state (a) and the input is a one then the encoder will always transition to state (b). Using the trellis diagram we can track the state changes in the encoder for any given input sequence.

The labels on the branches of the decoder trellis indicate the Hamming dis-

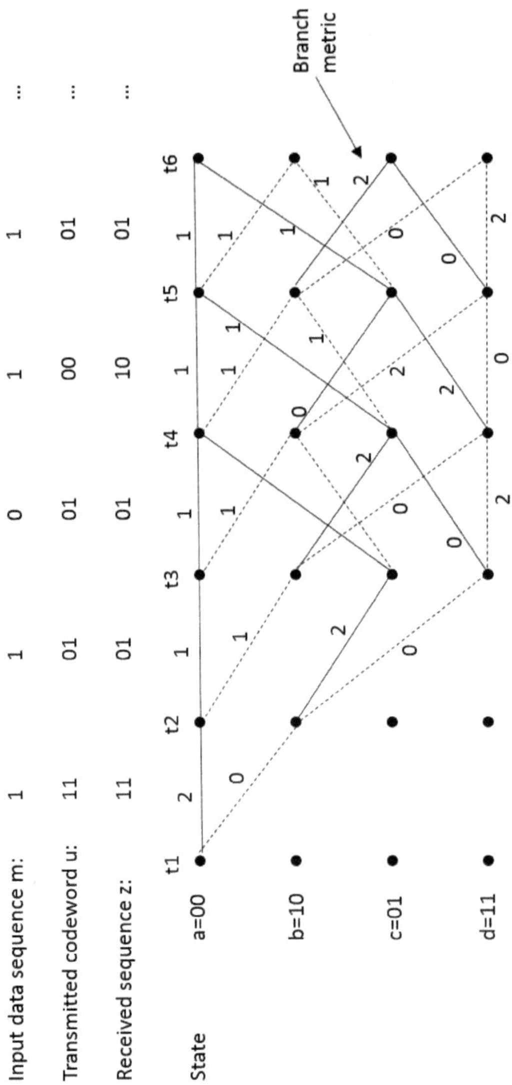

Figure 2.4.7: Decoder trellis diagram

tance between what has been received and the output that would have been generated at the encoder if the corresponding state transition had occurred. For example, at time t_1 we are in state $a = 00$. If the encoder had transitioned from $00 \to 00$ then the encoder output would have been 00, but the actual received codeword was $\mathbf{z} = 11$. Therefore, we label the transition from $00 \to 00$ with the Hamming distance between 00 and 11, which is 2. If the encoder had transitioned from $00 \to 10$ then the encoder output would have been 11, which is exactly what we received. Therefore, we label the transition from $00 \to 10$ with the Hamming distance between 11 and 11, which is 0. We continue to label the trellis this way as the symbols are received at each time t_i.

Clearly we are creating a set of divergent paths through the trellis, but the basis of Viterbi decoding is that if any two paths subsequently merge into a single state then one of them can always be eliminated as the *optimum path.* This is because of the Markov nature of the encoder which tells us that the present state summarises the encoder history, in that previous states cannot influence future states or future output branches. The two paths are effectively one path from this point onwards in the trellis, and therefore the merged path which has the larger cumulative Hamming distance is discarded because it cannot be the optimum path.

Example. Consider the two paths through the trellis illustrated in figure 2.4.8. The paths diverge at time t_2 but converge again at time t_5 into state $a = 00$. From that point on in the trellis the paths become the same, so the decision as to which of the two should be retained for further examination depends only on the metric penalty of each path prior to their convergence at time t_5. In this example the upper path has a Hamming distance of four and the lower path has a Hamming distance of one, so the upper path is discarded as being less likely to be the *maximum likelihood* solution.

References

[1] George F. Elmasry. *Tactical Wireless Communications and Networks: Design Concepts and Challenges*. Wiley, 2012.

[2] R.E. Larson and B.H. Edwards. *Elementary Linear Algebra*. D.C. Heath and Company, 1988.

[3] H. Nyquist. Certain topics of telegraph transmission theory. *Transac-*

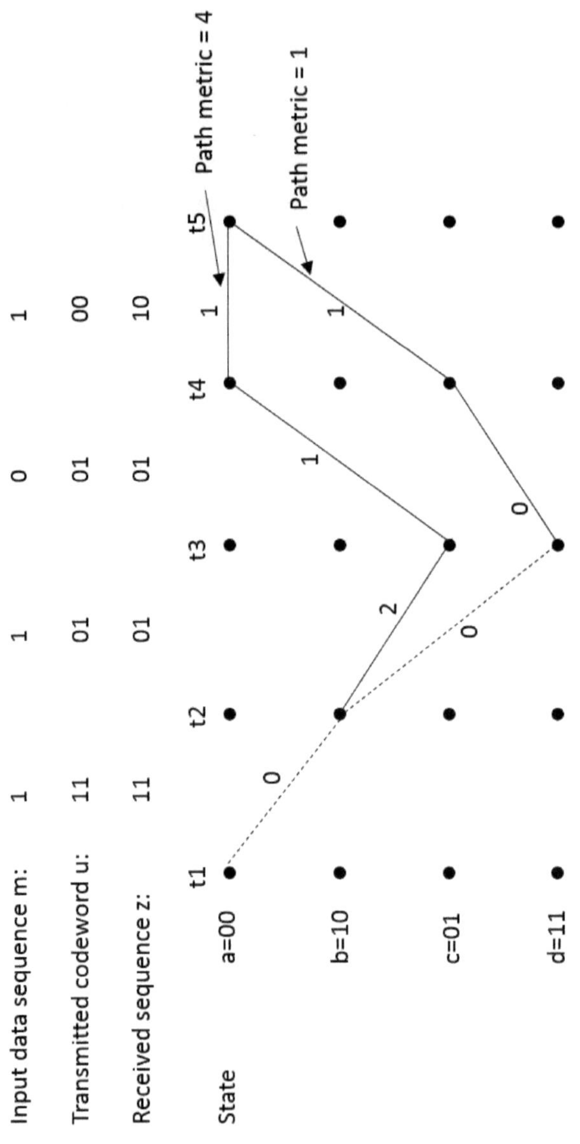

Figure 2.4.8: Path metrics for two merging paths

tions of the American Institute of Electrical Engineering, 47:617–644, April 1928.

[4] M. Ryan and M. Frater. *Battlefield Communications Systems.* Argos Press, 2007.

[5] Bernard Sklar. Reed-solomon codes, 2002.

[6] Bernard Sklar. *Digital Communications: Fundamentals and Applications*. Prentice-Hall, 1988.

[7] A. Viterbi. Error bounds for convolutional codes and an asymtotically optimum decoding algorithm. *IEEE Transactions on Information Theory*, IT13:260–269, April 1967.

[8] A.D. Whalen. *Detection of Signals in Noise*. Academic Press, Inc, 1971.

3 Legacy Tactical Data Architectures

3.1 Introduction

A communication architecture is the collection of people, systems and procedures that are used to transmit and receive information in order to achieve a set of operational outcomes. An architecture describes the entire problem space, but can be examined from several different perspectives. Operational views of an architecture show what the end-users are trying to achieve: who talks to whom, how often they communicate, what sort of information they exchange, what priority the information has, the required level of security and the distances between participants. The views will show intelligence exchanges such as maps, video and photographs. They will also describe the required sensor-to-shooter links and the command and control exchanges, including documents, verbal orders and video conference.

System views of the architecture show the underlying technology, at a system level, that will be used to support the operational requirements. Such views will illustrate the types of systems required and the interfaces between the systems, so that an end user's information requirement can be translated into a system-level data requirement.

Finally, technical views of the architecture show the specific technical specifications that underpin the systems in the architecture, such as data protocols, timing protocols and security protocols. The relationships between the various views are illustrated in figure 3.1.2. For a C4I system the operational needs are centred around command and control, and situational awareness. We will examine some of the system-level and technical-level solutions to those operational needs.

The system and technical design of a radio communications architecture needs to be built on a clear understanding of the technology characteristics described in chapter 2. Traditionally, military forces have used a system-level tactical communications design composed of two principal subsystems: the tactical

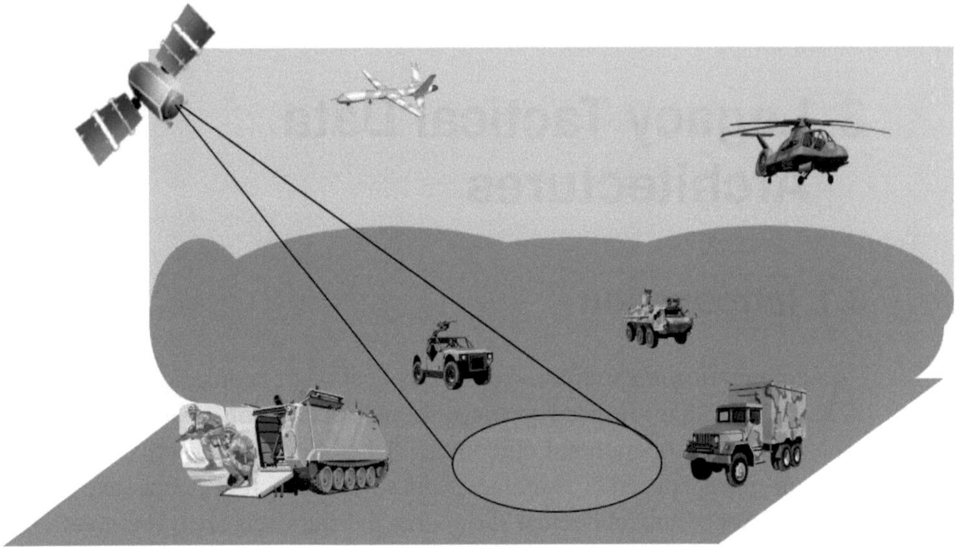

Figure 3.1.1: Operational views describe the end-user's perspective

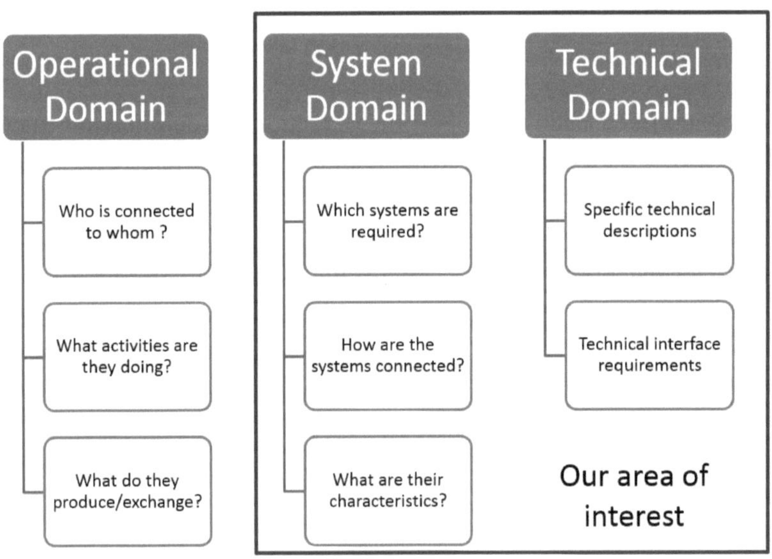

Figure 3.1.2: Architectural views

trunk communications subsystem and the tactical combat net radio (CNR) subsystem. Recent developments in tactical data communications, such as the deployment of wideband radios below battalion, are challenging this approach, but for most armies in the world this remains the effective organising principle. Generally, the trunk subsystem provides long range communications with moderate to high data rates, but lacks extensive tactical mobility. By contrast, the CNR subsystem has excellent mobility but only low data rates over short to medium distances.

The trend in modern battlefield communications is towards a homogeneous IP network that seamlessly connects all deployed forces and interfaces to one another and to strategic networks, providing the right capacity to the right place at the right time. However, as discussed extensively in [11], the core requirements of capacity, range and mobility in tactical communications compete against each other. For example, tactical elements need high capacity in order to receive the necessary situational awareness data, but that requires large, high gain, directional antennas. Mounting such antennas on a tactical vehicle impedes the vehicle's mobility, which is essential for survival on the modern battlefield. Operating in dispersed formations increases the distance between force elements, but maintaining communications then requires long mast antennas or to seek high ground. Long mast antennas again reduce mobility and seeking the high ground for communications conflicts with basic tactical procedures for cover and concealment. Some communications technologies can provide high capacity over short range; some can provide low capacity but good mobility and some can achieve long range but only moderate capacity and reduced mobility.

Currently, no single system can meet all of the tactical requirements shown in figure 3.1.3. The US Joint Tactical Radio System (JTRS) program was an attempt to produce a single solution by creating a family of software defined radios that could be reconfigured to service the full range of activities required for tactical communications. After a decade of effort and billions of dollars the program was effectively closed in 2012, and reference [3] provides an interesting analysis of some of the problems faced by the program. For the foreseeable future, although a single logical network may be achieved, the physical network is likely to continue to comprise multiple technologies and systems. Many countries are in a hybrid position in which they continue to maintain older generation systems, but implement upgrades of those systems to provide necessary data services. This is largely due to the enormous cost involved in military procurement and the long time frames required. For

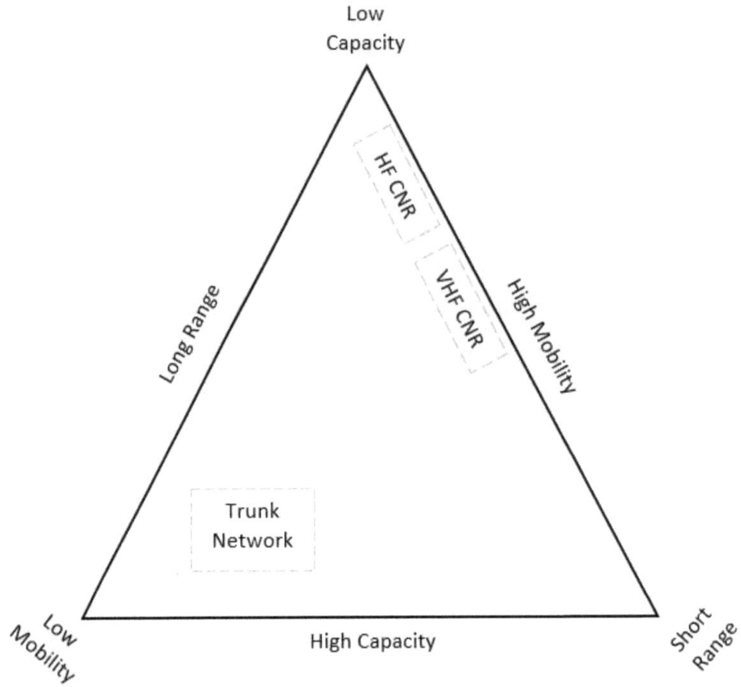

Figure 3.1.3: Range-Capacity-Mobility trade off [11]

many countries the process of purchasing and fielding an entirely new military communications system is a ten or fifteen year activity which can have the unfortunate consequence of delivering the last generation of technology, rather than the current generation.

The characteristics and limitations of trunk radio and CNR will continue to influence the effectiveness of tactical data networks. For this reason it is important to understand what this existing generation of military communications looks like and what it was designed to do. Once that is understood, the size of the technological change that is occurring from the last generation to the current generation becomes clearer. For acquisition professionals the issue to be resolved is at what point should the older system be retired, rather than upgraded, and if a new system is to be purchased then how might its currency be ensured against the long military acquisition timetable.

3.2 History

Figure 3.2.1 illustrates the timing of some of the significant advances in tactical communications over the last thirty years. In the mid-1980s the major armies around the world each commenced the process of digitising the battlefield. Up until that point all of the major tactical trunk systems were analogue systems, but commercial experience had shown that digital technology could offer much more flexibility and security to tactical communications. Digital signals can be subjected to rigorous cryptography for protection and in the commercial world digitisation was providing better integration of voice and data services.

Major military acquisition programs were initiated to develop digital trunk communication systems. In the UK the Ptarmigan program would replace the ageing BRUIN system with a new corps level trunking system. In the US the Mobile Subscriber Equipment (MSE) program would introduce a fourth generation mesh topology for tactical trunk communications. In Australia Project Parakeet was conceived to bring a new integrated digital voice and data system to the Australian Army, replacing the aged AN/MRC 127 radio relay. The core technologies in each of these projects was similar, despite their separate development paths.

Fourth generation trunk systems were still heavily voice oriented, with their principal purpose being to provide telephone trunk connections between nodes on the battlefield. Therefore, the systems of the 1980s and early 1990s provided circuit switched 16 kbps channels, which were multiplexed together for transmission as a 2 Mbps signal over a point-to-point bearer. Voice channels were encoded using 16 kbps Continuous Variable Slope Delta Modulation (CVSD) which could provide a reasonable quality telephone channel in one quarter of the bandwidth used for digital voice in commercial systems, and each conversation occupied a single circuit. Data was carried using a secondary packet switch, with the circuit switch providing the physical layer for the data network. It was possible to aggregate several of the 16 kbps channels in order to increase the data capacity, but such arrangements were quite inflexible. The native protocol supported in these trunk systems was X.25, which was a popular connection-oriented data protocol. It is suitable for carrying low data rate traffic that is not delay sensitive, such as text-based messaging, but it is not suitable for high-rate services.

During the 1990s two events occurred that showed up the weaknesses in these relatively new tactical trunk systems. The first was the Gulf War in 1991, which saw coalition forces moving rapidly across large distances. Both the US and UK forces discovered that the mobility of their trunk systems was not adequate and they could not keep up with the force they were supposed to be supporting. Ptarmigan and MSE had been developed against a Cold War threat in Western Europe, but the dispersion of forces during the Gulf War required communications between nodes that were separated by large distances. Ad-hoc adjustments were made to improvise the communications network, particularly by using satellite trunks to connect tactical nodes. The second weakness was the extended time it took to set up the tactical trunk infrastructure at each headquarters. In a Cold War scenario it was anticipated that a brigade headquarters would prop for six hours [7], providing some lead time to set up the radio relay chain and lay out the headquarters infrastructure. But during the Gulf War the formation headquarters moved far more rapidly, such that they were often moving again before the trunk communications had even been established. Both of these lessons seem to have influenced the US approach to tactical communications during the subsequent twenty years. The US forces have prioritised communications on-the-move and between dispersed forces in their development program.

The second major event in the 1990s that influenced tactical communications was the rise of the Internet. Although the original network, developed by DARPA, had been in existence for many years, it was the development of the *hypertext transport protocol* in the early 1990s that opened up the Internet to the general public. Web servers and browsers offered a simple, intuitive interface for the average end-user and the possibilities quickly expanded. Military commanders using email and an intranet in barracks began to expect similar services when deployed. The problem, apart from a lack of bandwidth, was that the major tactical trunk systems such as Ptarmigan, MSE and Parakeet were connection oriented, including the modest X.25 data component, whereas the Internet applications are built on the connectionless TCP/IP protocol suite. There is still debate about whether TCP/IP is an appropriate protocol for wireless tactical communications because it introduces significant overhead bits which impairs network efficiency, particularly for short payloads. However, the protocol suite has become so ubiquitous that it has become the de-facto standard for network communications.

Efforts to take internet-like services to the field grew during the 1990s with the US eventually deploying the Tactical Internet. The US Non-classified Internet Protocol (IP) Router Network (NIPRNET) and the Secret Internet Protocol (IP) Router Network (SIPRNET) were established by creating IP gateways between the MSE trunk network, the EPLRS situational awareness network and the SINCGARS CNR network. None of these systems were natively IP-based systems, but each was modified to support the TCP/IP protocol suite by creating IP gateways between the networks.

In 2001 the terrorist attack in New York commenced a decade of war in Iraq and Afghanistan which brought the communication needs of military forces into sharp focus. In the US, the Joint Tactical Radio System (JTRS) program initiated a bold move to harness advanced digital signal processing (DSP) technology to create a family of software defined radios for all US tactical forces. At the same time, the Warfighter Information Network - Tactical (WIN-T) program commenced with the aim to incrementally replace the MSE trunk system with a wideband native IP data network. In the UK a replacement was sought for the ageing Ptarmigan system and in Australia Joint Project 2072 promised a complete overhaul of tactical communications. None of these activities have been smooth travelling, with the JTRS program halted in 2011. Whilst good work was done in standardising future tactical waveforms, the dream of a software defined tactical radio was not realised. Nevertheless, during that time several commercial companies have deployed new products into Iraq and Afghanistan that have achieved significant results in providing high speed data on-the-move. In the UK the new Falcon trunk system is being delivered and the Bowman digital CNR has been introduced into service, although apparently with some problems.

3.3 Non-IP Tactical Data Networks

The hierarchical, connection oriented, voice-based communication systems available in the 1980s were not really suitable as data networks. Telephone connections were available within the trunk network and voice connections were possible from the CNR system into the trunk communications system using a suitable CNR interface, but data transfer capacity was inadequate. The simple example given in section 1.3.2 illustrates how the data capacity available in the system simply did not match the information flows required to prosecute the battle. The hierarchical nature of CNR nets created data bottlenecks that inhibited the transfer of situational awareness data.

As early as the late 1970s it was recognised that sharing situational awareness data was not easily achieved using conventional circuit switched communication systems. Typically the holder of the data did not know exactly who

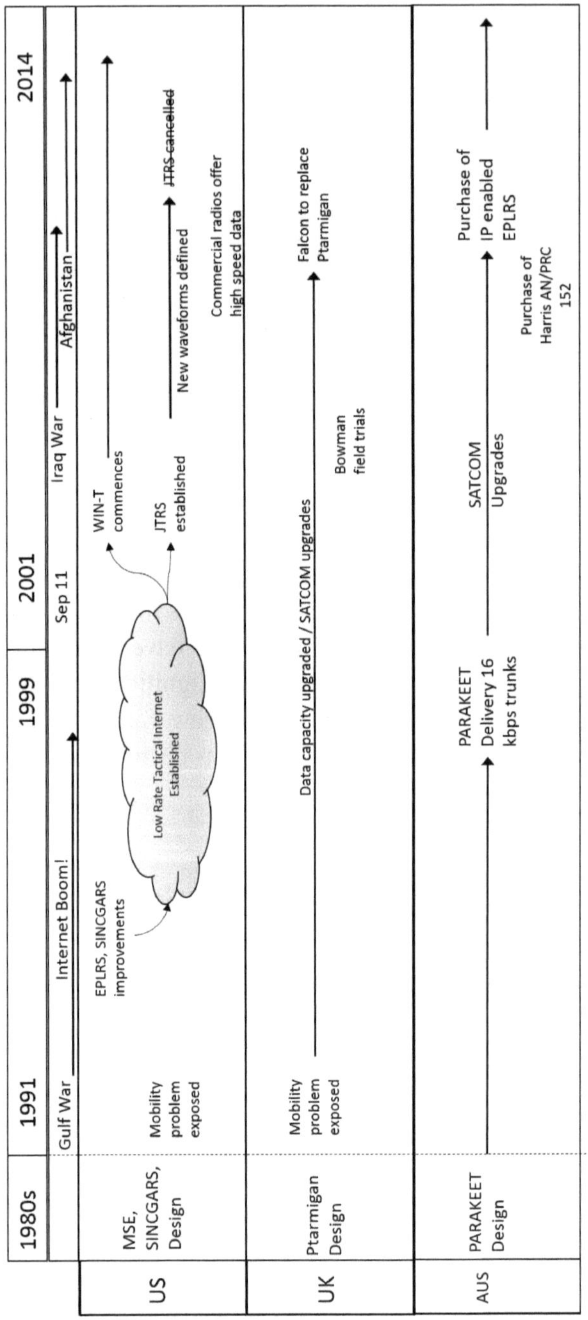

Figure 3.2.1: Military technology milestones

wanted the information and a data seeker often did not know who to ask for a particular piece of information. A new approach to data sharing was required in order to provide comprehensive situational awareness data to all of the relevant force elements in a theatre of operations.

In order to share data in a systematic way certain design decisions had to be taken in the development of new radios and ancillary data terminals. Essentially, the designers had to package together the various technologies discussed in chapter 2 in such a combination as to optimise the reliable transfer of situational awareness information from sources to sinks. In the tactical domain the package of technologies when taken together is referred to as a *waveform*. A waveform is the entire set of radio and/or communication functions that occur from the user input to the radio frequency output and vice versa. There have been two significant non-IP waveforms that have provided effective tactical data networks in support of situational awareness: Link 16 and EPLRS. Both networks are still widely used by numerous military forces.

3.3.1 Link 16/JTIDS

The Joint Tactical Information Distribution System (JTIDS) is one of the most sophisticated wireless tactical data systems in use by US, NATO and coalition forces around the world. It supports time sensitive, robust radio communications and has been integrated into many platforms, primarily in the air and maritime domain. The underlying waveform on which JTIDS is built is Link 16. Link 16 supports voice, free text and variable format messages. It is a complex waveform that uses TDMA, FDMA and CDMA access schemes. Because the waveform was developed well before the protocol stack in figure 2.1.1 was formalised, it approaches network performance from a different perspective. The core functions of the waveform are in the physical and data link layers; the MAC layer function of managing access to physical resources is performed during the planning phase by the allocation of time slots in the TDMA architecture, and a 75-bit block of data, called a J-word, is the equivalent of a MAC frame or IP packet.

Link 16 uses FDMA to separate the spectrum into 127 possible nets. For tactical robustness Link 16 uses frequency hopping spread spectrum techniques on each of the 127 nets to minimise susceptibility to electronic attack. Each net uses a unique frequency hopping code which is applied to every terminal in a given net. The pseudo-random hopping codes are orthogonal to ensure minimum interference across nets that are occupying the same total bandwidth.

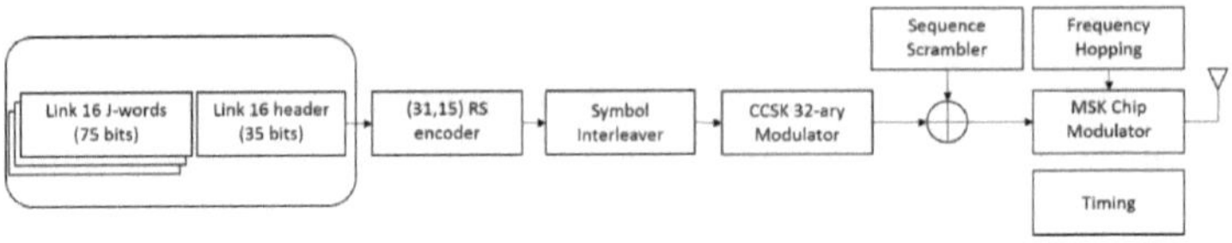

Figure 3.3.1: Link 16 Block Diagram

A Link 16 net hops around 77,000 times per second into one of 51 frequencies in the bandwidth between 960 MHz and 1215 MHz, with each frequency separated by 3 MHz.

Each net established in a Link 16 environment is managed with a TDMA access scheme. Each TDMA "epoch" is 768 seconds (12.8 minutes), comprised of 64 frames. Each timeslot in the frame is 7.8125 ms in duration, which creates a total of 1,536 timeslots in a frame, or 98,304 timeslots in each epoch. For management, the timeslots are designated into groups A, B and C such that slots can be numbered A0, B0, C0 .... A32767, B32767, C32767. Within its allocated timeslot a terminal transmits a series of 6.4 μs pulses, each separated by an interval of 6.6 μs. Each pulse includes a data stream that has been spread in frequency by using a DSSS method, as described in section 2.2.7, to provide further resilience against electronic attack. In Link 16 fast hopping occurs every 13 μs, approximately 600 times per timeslot.

Link 16 terminals are pre-assigned to a set of timeslots and the data rates available are 28.8, 57.6, 115.2 and 238 kbps [5]. Voice is supported through two 16 kbps channels with a push-to-talk protocol. Error coding is configurable, but includes the following modes:

- parity based error detection,
- Reed-Solomon coding for error detection and correction,
- symbol interleaving to randomise the effects of burst errors, and
- symbol repetition.

Link 16 Message Format

Figure 3.3.1 shows the various components that form the Link 16 waveform. A Link 16 message comprises a 35 bit header and a data part. For tactical messages the data part is transmitted using a fixed format called a J-Series message. In a J-Series message there are three types of J-Series words: initial, continuation and extension, with each word being 75 bits long. Of these

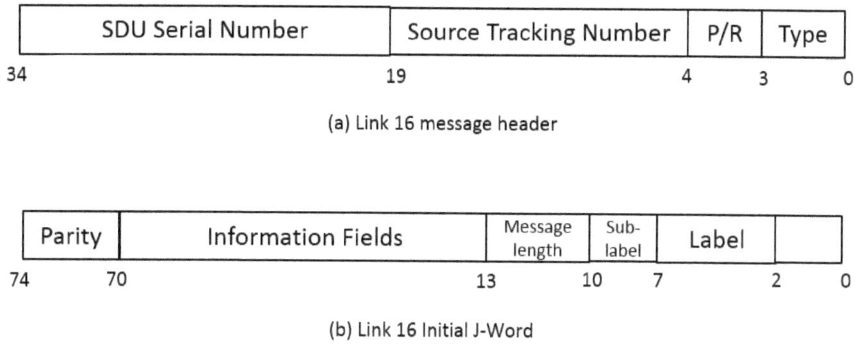

Figure 3.3.2: Link 16 Message Format

bits, the first 70 bits are information and four bits are parity bits, with one bit reserved. J-Series words can be packed into the timeslot pulse in several ways, depending on the electronic threat and noise conditions. The standard "double-pulse" packing structure is to transmit three J-Series words in one timeslot (225 bits), but if the Pack-4 structure is used then twelve words are transmitted in each slot. Messages are taken in groups of three, six or twelve J-words. The header and initial J-word format are shown in figure 3.3.2.

Link Layer Operations

The link layer assembles the words for transmission prior to passing them to the physical layer. Each timeslot transmits the 35 bit header to describe the data payload, so the link layer records the transmit terminal's packing structure in the third bit of the header. The twelve parity bits are generated using a binary (237,225) cyclic redundancy check error detection code. and then distributed across the four parity bits (position 71-74) in each J-word. Finally, the SDU serial number is encrypted and stored in the header at bits 19 to 34. The assembled package is passed to the physical layer, where Link 16 uses Reed-Solomon coding, symbol interleaving and cyclic code shift keying (CCSK) for M-ary baseband symbol modulation. Minimum Shift Keying (MSK) is used for transmission, including double pulse diversity and a combination of frequency hopping and direct sequence spread spectrum for electronic resilience.

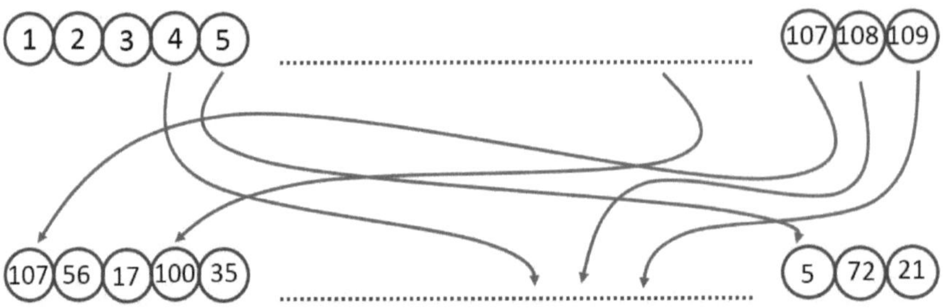

Figure 3.3.3: Symbol interleaving for burst resilience

Reed-Solomon Coding

Link 16 uses a (31, 15) Reed-Solomon code, where each symbol is five bits ($m = 5$). This code can correct $t = \frac{n-k}{2} = \frac{31-15}{2} = 8$ symbols in each 31 symbol codeword, or in other words, a burst error of up to 40 bits [13]. When the Link 16 message arrives at the physical layer it is mapped onto five-bit symbols. For example, the 35 bit header becomes seven symbols and each 75 bit J-word becomes fifteen symbols. Figure 2.4.3 illustrates this concept. The header is encoded using a RS(16,7) code, which is related to the (31,15) code applied to the message frames. After coding, the header is represented by sixteen symbols and each J-word is represented by thirty one symbols. This means that the coding for the header is less robust than the coding for the message and this is why the header is always transmitted using a double pulse.

Symbol Interleaving

A standard Link 16 package consists of a header followed by three J-words, which consists of sixteen symbols for the header, followed by thirty one symbols for each of the three J-words. Hence there are a total of 109 symbols to be transmitted. The symbol interleaver is primed with a pseudo-random sequence to select the 109 symbols for transmission in a random order, as illustrated in figure 3.3.3.

The benefits of symbol interleaving are twofold: firstly, separating the symbols in a J-word makes the transmission less susceptible to a burst transmission error. A burst will affect a consecutive series of symbols, but with symbol interleaving each symbol is likely to be from a different part of the Link 16 message. For example, ten consecutive symbols might be corrupted, but after

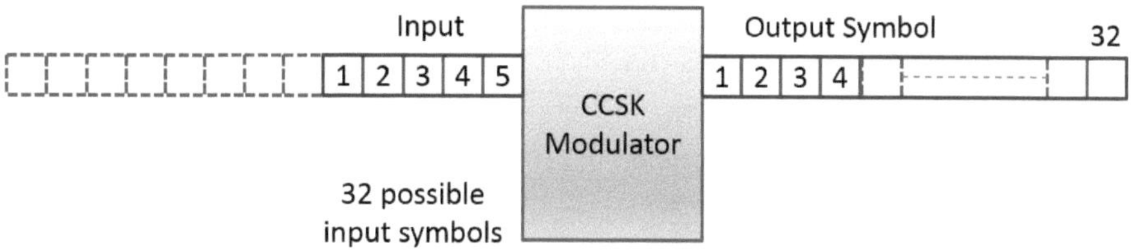

Figure 3.3.4: CCSK modulation

de-interleaving at the receiver they are returned to each of the three J-words. Thus, each J-word now only needs to correct two or three corrupted symbols, rather than ten symbols. The second advantage of interleaving is that it introduces a first layer of traffic security.

CCSK Baseband Symbol Modulation

After interleaving has occurred a further thirty two synchronisation symbols and eight time refinement symbols are appended to the front of the interleaved, coded symbols. Baseband modulation is then performed using cyclic code shift keying (CCSK), in which each five bit symbol is replaced with a thirty two bit chip sequence [5]. The chip sequences are derived from a base sequence, which is then cyclically shifted by one bit to the left a further thirty one times to create the library of thirty two different chip sequences. Clearly, the chip rate is $\frac{32}{5}$ times faster than the pre-modulated bit rate, which has the effect of spreading the bandwidth of the signal. After the CCSK modulation each thirty two bit chip sequence is scrambled with a thirty two bit pseudo-random sequence which provides a second layer of transmission security and ensures even spreading of the signal energy across the frequency band. The process is illustrated in figure 3.3.4.

Physical Layer Operations

Minimum Shift Keying (MSK)

The chip sequence is modulated onto a carrier using Minimum Shift Keying (MSK). MSK is a form of Continuous Phase Modulation (CPM) in which sharp phase transitions are avoided by changing the phase across the full bit period. MSK can be considered as a special form of orthogonal QPSK, with

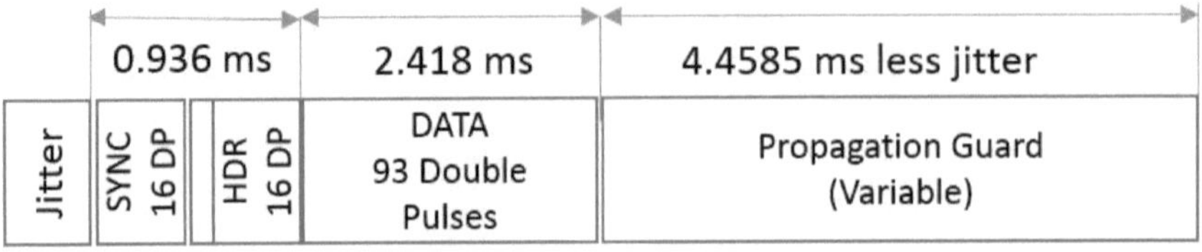

Figure 3.3.5: Standard double pulse packing

sinusoidal symbol weighting, or a special case of continuous phase FSK [14]. Considered as the latter, for binary chips ± 1 the two transmitted signals are

$$
\begin{aligned}
s_1(t) &= \cos\left(2\pi\left(f_c + \frac{1}{4T_s}\right)t\right) \\
s_2(t) &= \cos\left(2\pi\left(f_c - \frac{1}{4T_s}\right)t + \pi\right),
\end{aligned}
$$

where T is the chip period and f_c is the carrier frequency. Over a chip period $s_1(t)$ shifts the phase of the signal, relative to the carrier, f_c, positive $\frac{\pi}{2}$ radians and $s_2(t)$ shifts the phase negative $\frac{\pi}{2}$ radians. The nett effect is that the signal avoids sharp phase transitions, which reduces out-of-band signalling, and the signal has a constant envelope, which aids power performance. Note that Link 16 supports message data being sent as a single pulse and also a double pulse, depending upon the packing structure. The double pulse carries the same information, but on two different frequencies. Clearly this will improve the error performance of the underlying data against fading or narrow band jamming, but at half the data rate [5]. The standard double pulse packing structure is shown in figure 3.3.5.

When this standard structure is used the Link 16 terminal transmits three J-words per slot. Since each J-word is 75 bits, and the Reed-Solomon code uses five bit symbols, each J-word is 15 symbols, which becomes 31 symbols after error coding. After CCSK encoding the J-word is represented by 31 symbols, each of 32 bit size, or a total of 992 bits. Encoding three J-words per slot means the transmission of 992x3 = 2976 bits in 93 pulses, or 32 bits per pulse.

The pulse structure is $6.4\mu s$ of transmission, followed by $6.6\mu s$ of dead time, so the chip rate is 32bits/$6.4\mu s$ = 5 Mbps.

The jitter component is a pseudo-random delay prior to transmission, which enhances the security of the system by making it more difficult for a following jammer. The propagation delay is dead time to avoid collision with the next transmission. The minimum delay time is 2 ms, which can support a network dispersed over 315 nm.

Conclusion

It is hard to imagine that the Link 16 waveform was developed in the 1970s, given its incredible level of robustness and its capability in reliably connecting jet fighter aircraft over 500 km apart travelling faster than sound in a jamming environment. Link 16 was the first tactical waveform to combine frequency hopping, direct sequence spread spectrum, concatenated codes and TDMA. The more recent challenge has been to integrate Link 16 nets into the emerging all-IP tactical core network with suitable gateways.

3.3.2 Enhanced Position Location and Reporting System (EPLRS)

Background

The original Position Location Reporting System (PLRS) was developed towards the end of the Vietnam war to reduce fratricide among land forces. It was developed by the US Marine Corps, but then became a joint Army/Marine program beginning full production in 1983. The PLRS mission was precise: to provide units with their own location data whilst maintaining situational awareness at a PLRS Master Station located at the tactical level headquarters [4]. Determining the positions of individual units was achieved through a fairly complex system that involved establishing fixed (stationary) reference points within the deployed architecture, exchanging time-stamped data between multiple nodes, and then comparing the time of arrival (TOA) of the exchanged data with its time stamp when received at the receiving nodes. These TOA, coupled with altitude estimations determined by barometric pressure and temperature readings taken at each individual radio, were exchanged throughout the nodal architecture to calculate accurate geographical positions for each unit. These calculations were based on differential triangulation and converted to the military grid reference system (MGRS). PLRS was bulky, expensive and complex but it worked.

With the introduction of the Global Positioning System (GPS) the position location function of PLRS became largely redundant, providing at best a sec-

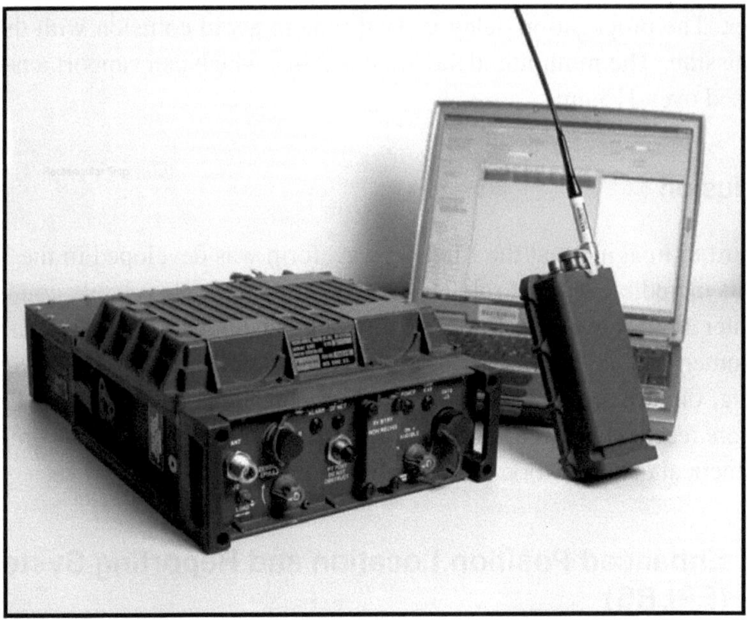

Figure 3.3.6: EPLRS Radio and EPLRS Microlight

ondary system should GPS become unavailable. However, it had been fielded to transmit data messages across the battlefield and as the need for tactical data was growing it evolved, through a program of enhancements and upgrades, into a genuine tactical data radio. With the rise of Network Centric Warfare in the 1990s, the Enhanced PLRS (EPLRS) became the backbone network for the US tactical internet down to battalion level and it has been taken up by several other military forces around the world. It can provide long range, non-directional broadcast communications with automatic data relay, which makes it a viable tactical Mobile Ad-hoc Networking (MANET) solution.

EPLRS is manufactured by Raytheon (see figure 3.3.6) and was recently demonstrated in Afghanistan as a wideband, IP-based tactical internet backbone that can provide data on-the-move to military vehicles [10]. The EXF1915, an upgraded version of the EPLRS radio, completed several months of continual, sustained secure data transmissions for the combat vehicles of the 4th Brigade 2nd Infantry Division Stryker Brigade Combat Team, or 4/2 SBCT. Soldiers were able to send and receive e-mail and chat messages and access the brigade's intranet-like Web portal, marking the first time 4/2 SBCT was able to tap into a secure wireless network on-the-move [10].

EPLRS Operation

The EPLRS network consists of many EPLRS Radio Sets (RS) and one or more EPLRS Network Manager (ENM) host computers. The US operational concept for EPLRS is for it to provide tactical data from Regimental HQ down to Battalion or Company HQ, as shown in figure 3.3.7. It also provides an interface into the lower level SINCGARS tactical data network, achieving tactical data connectivity from the forward elements to the Tactical Edge.

EPLRS is a self-forming and self-healing system, so it can cope with stations appearing and disappearing from the network. Routing is established when a message is transmitted, passing through whichever other EPLRS radio stations are within communications range of the originator. Therefore, the originator and the receiver do not need to be in direct line of sight, as they do in hierarchical voice nets.

EPLRS Needlines

Each RS automatically routes and delivers user messages, as directed by the ENM, and provides multiple concurrent communication paths known as *needlines* [15]. At each RS the host computer may be an ENM or another type of host device.

The TDMA slot structure is used to set up virtual circuits between hosts across the EPLRS network. These virtual circuits, or needlines, can be point-to-point or broadcast services. A single RS host can support 32 concurrent needlines which allows it to send and receive information from different groups of hosts, although typically only 28 are available because the remainder are required for network coordination [6]. There are four types of needlines:

Carrier-Sense Multiple Access (CSMA). CSMA needlines provide the capability for many hosts to communicate with one another to share situational awareness or command and control data. CSMA needlines operate like the Ethernet protocol, or a voice contention radio net. Each RS senses whether the virtual circuit is in use and when it is clear then the RS transmits its data. All other RS on the needline receive the data.

Multi-Source Group. The MSG needline is a few-to-many broadcast needline in which messages are transmitted by a select few RS source endpoints and then received by a large number of RS sink endpoints. An MSG needline can have up to sixteen source RS with data rates up to 488 kbps. They are

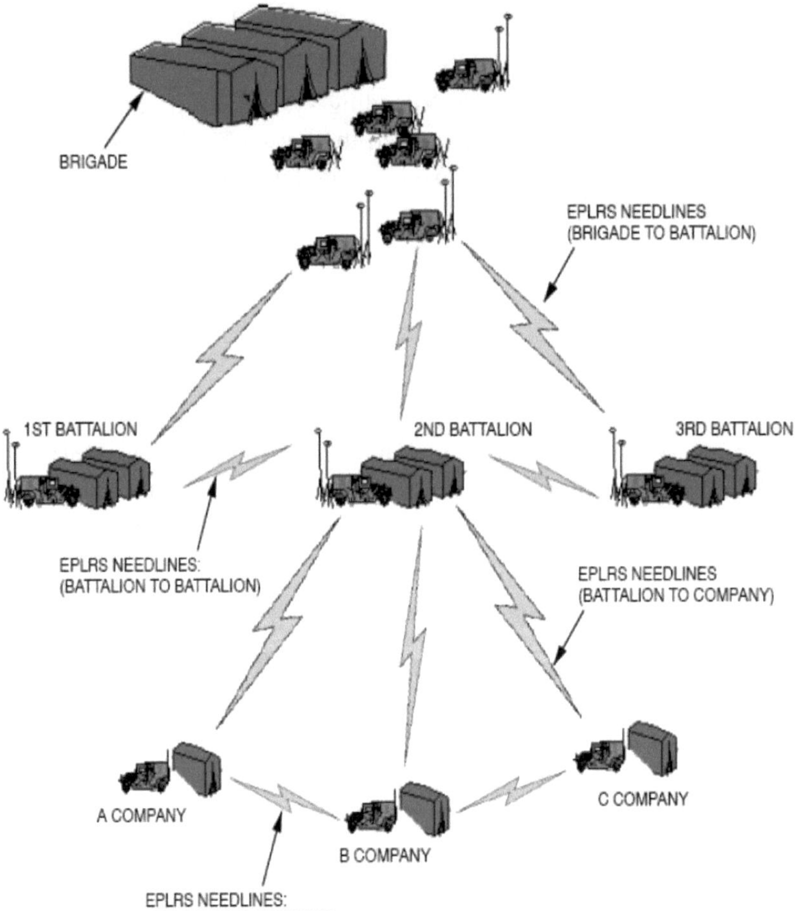

Figure 3.3.7: EPLRS Concept [1]

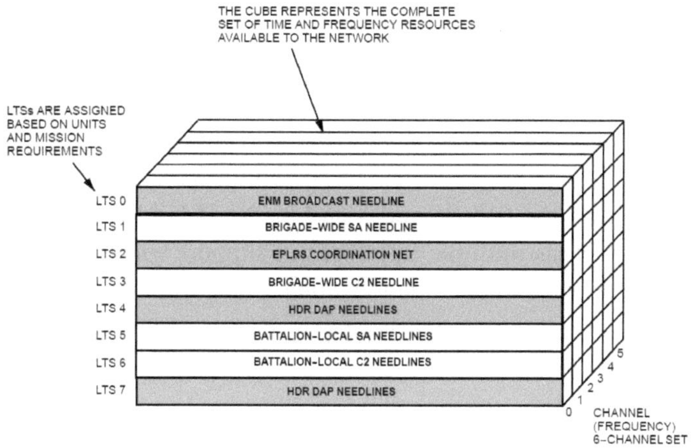

Figure 3.3.8: Example of EPLRS resource allocation [1]

typically used for broadcasting low latency track data, movement orders or data exchange.

High Data Rate (HDR) Duplex. HDR duplex needlines are point-to-point circuits that provide acknowledged communications between the hosts. The circuit is duplex in that each end RS shares an equal number of timeslots on an alternating basis. Data rates are up to 240 kbps each way and they can be used for data intensive messages, such as video and large file transfers. The coordination network automatically selects a path through the network by assigning specific relay nodes to establish the links that connect the endpoints. This process involves the assignment of frequency and time resources to each node in the needline to guarantee available bandwidth between the endpoint RS.

Low Data Rate (LDR) Duplex. These needlines are similar to HDR duplex except that they offer lower data rates each way, typically from 20 bps to 16 kbps. LDR needlines are utilised for Voice over IP (VoIP) type communications.

A typical allocation of needlines to Logical Timeslots and frequency channels is shown in figure 3.3.8.

EPLRS Multiple Access

User-to-user data communications are provided by the high security anti-jamming EPLRS waveform. A single EPLRS net can be reconfigured as a Local Area Network (LAN), a Wide Area Network (WAN), or range extension to provide a greater area of coverage within the area of operations [5]. Figure 3.3.6 shows the EPLRS radio and the hand held EPLRS Microlight. The EPLRS radio has multiple interfaces to support voice communications, data communications and position location reporting.

TDMA. Like Link 16, the fundamental access scheme for the EPLRS network is TDMA. A 64 second epoch is broken into 256 frames of 0.25 seconds, with each frame comprising 128 time slots, as illustrated in figure 3.3.9(a). Each time slot is approximately two milliseconds and 16 timeslots are grouped together into eight logical timeslots as shown in figure 3.3.9(b). Transmission takes 800-1000 microseconds and the remaining time is required for processing overhead and message validation [5]. An alternate configuration supports timeslots with a four millisecond duration, but this choice must be made when planning the deployment [4].

FDMA. Channel re-use within a geographic area is achieved by using eight channels across the UHF spectrum from 420 - 450 MHz. Each channel is 3 MHz wide [11]. EPLRS also has a frequency hopping mode in which the signal hops across eight frequencies at a hopping rate of 512 hops per second. When two millisecond timeslots are used this arrangement transmits each timeslot on a different frequency.

CDMA. Similar to Link 16, the EPLRS transmission is also encoded using a DSSS modulator with a pseudo-random chip sequence at a rate of 5 Mcps (5×10^6Chips per second). This provides robustness against narrow band jamming through the spreading gain, which is around 8 dB, but also helps to manage inter-net interference. Inter-net interference on an EPLRS network arises because there are only eight frequencies on which the system can hop, which can result in frequency clashes. By using CDMA each needline can be separated by code to ensure safe transmission.

EPLRS Data Link Layer

Similarly to Link 16, EPLRS uses Reed-Solomon encoding, with interleaving, to provide an error correcting capability and a resilience against burst interference. Because synchronisation is required for the TDMA process, the

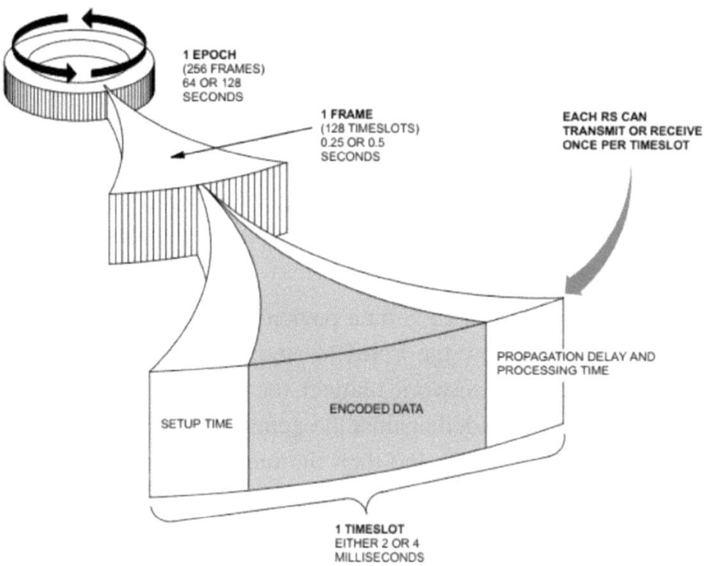

(a)

LTS0	0	8	16	24	32	40	48	56	64	72	80	88	96	104	112	120
LTS1	1	9	17	25	33	41	49	57	65	73	81	89	97	105	113	121
LTS2	2	10	18	26	34	42	50	58	66	74	82	90	98	106	114	122
LTS3	3	11	19	27	35	43	51	59	67	75	83	91	99	107	115	123
LTS4	4	12	20	28	36	44	52	60	68	76	84	92	100	108	116	124
LTS5	5	13	21	29	37	45	53	61	69	77	85	93	101	109	117	125
LTS6	6	14	22	30	38	46	54	62	70	78	86	94	102	110	118	126
LTS7	7	15	23	31	39	47	55	63	71	79	87	95	103	111	119	127
	TSI0	TSI1	TSI2	TSI3	TSI4	TSI5	TSI6	TSI7	TSI8	TSI9	TSI10	TSI11	TSI12	TSI13	TSI14	TSI15
	<----------Frame---------->															

(b)

Figure 3.3.9: EPLRS TDMA Structure [1]

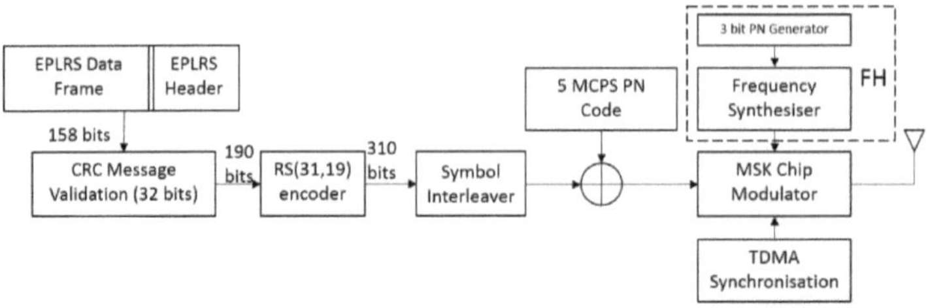

Figure 3.3.10: Block Diagram of EPLRS system

modulation scheme is Minimum Shift Keying which reduces out of band energy and maintains a constant signal envelope which aids in power management. Figure 3.3.10 provides an outline of how these component technologies combine to build the EPLRS system.

The exact structure of the EPLRS data packet shown in figure 3.3.10 is programmable in order to allow the ENM to adapt the system to the electronic threat. EPLRS offers the network planner the range of waveforms in table 3.1, where throughput is traded against the general anti-jam performance. For example, if *waveform 0* is selected then the underlying data rate is 38 kbps, which equates to approximately 80 bits transmitted in each 2 millisecond time slot. In the absence of any external interference, this waveform will successfully communicate with another RS at a distance of 110 km as long as the received signal power remains above -100 dBm for 90% of the burst transmission in the timeslot.

The Internet Protocol (IP) on EPLRS

The original EPLRS system, like most military data systems from the 1980s was designed for the X.25 protocol which provides connection oriented virtual circuits for low data rate messaging. EPLRS networks were effectively closed networks, because it was difficult to exchange data with other battlefield data networks. In a similar way, the US combat net radio subsystem, SINCGARS, which also had a basic data capability, was another closed network, which made tactical data exchange difficult.

In 1997 an Internet Controller (INC) was added to EPLRS to provide an Internet Protocol (IP) routing capability, as part of the development of the Tactical Internet. The INC acts as an IP gateway that links the EPLRS net with other networks, particularly battlefield trunk networks. Today, the INC is simply a

Waveform Group (Timeslot)	Waveform Mode	Max Data Rate (KBPS)	User Data Bytes per Trans-mission	General Anti-Jam Performance	dBm for 90% Burst Throughput	Maximum RS-to-RS Propagation Range (No Relays) (Km)
Tactical Internet (2ms)	0	38	10	Better	-100 (dBm)	110
	1	38	10	Best	-102 (dBm)	110
	2	77	20	Better	-100 (dBm)	110
	3	115	30	Good	-98 (dBm)	110
	4	311	81	OK	-94 (dBm)	110
	14	430	122	OK	-94 (dBm)	28
Expanded Data (4ms)	5	65	34	Best	-102 (dBm)	90
	6	127	66	Better	-100 (dBm)	90
	7	184	96	Good	-98 (dBm)	90
	8	230	124	Good	-98 (dBm)	90
	9	486	253	OK	-94 (dBm)	90
	17	772	402	OK	-93 (dBm)	35
	18	1014	528	OK	-92 (dBm)	23

Table 3.1: EPLRS Waveform Modes [1]

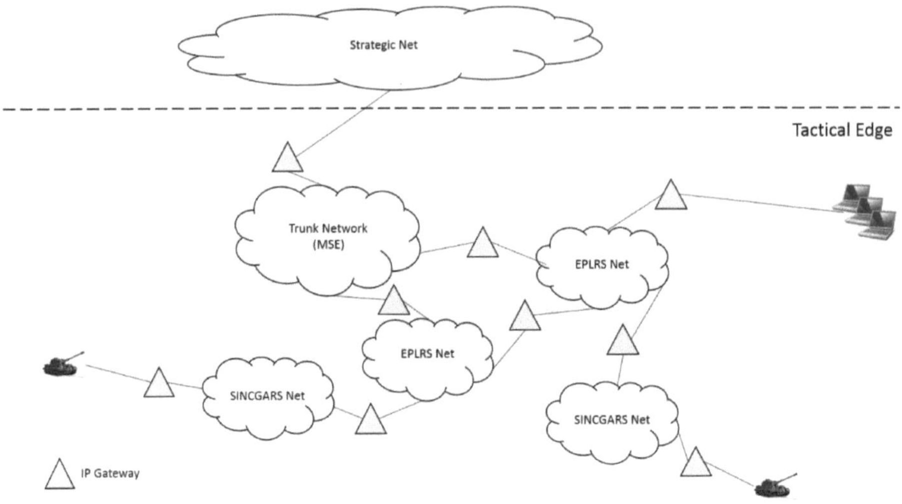

Figure 3.3.11: Integrated tactical networks [5]

circuit card in an EPLRS radio that is designated as the IP gateway. That radio maintains a map of IP addresses to each EPLRS RS in the network, similar to the Network Address Translation (NAT) performed by standard commercial IP routers.

The introduction of an IP gateway to the EPLRS network allowed the development of standard TCP/IP services across the Tactical Internet. The US military mandated that all network touch points, including to the lower level SINCGARS network, would be IP based, which then provided for seamless transition of data from the forward edge of the tactical domain to strategic networks, as illustrated in figure 3.3.11 [5].

EPLRS uses a flood routing approach that can respond to battlefield dynamics. Links can be dynamically reassigned when their reliability drops below an acceptable level, without relying on a central net control station. Dual path routing is also used to improve the network's robustness to sudden changes in jamming and to increase the percentage of needline satisfaction under dynamic jamming [12]. Whilst it is simple to set up and robust, flood routing can be inefficient in its use of resources. Consider an EPLRS network that includes a relay on a CSMA needline to some nodes that are geographically separated. If an entire logical timeslot is assigned to the needline, then without the relay there are sixteen possible transmission slots that the nodes can compete for. However, once a relay is introduced, one time slot must be dedicated to the relay function for every transmit slot, effectively halving the available bandwidth. If this is not done then the EPLRS packets will not propagate across the network and the broadcast assumption for the needline cannot be guaranteed. If the deployment of the EPLRS network demands two relays (three hops) then two dedicated relay slots are required for each transmit slot, effectively reducing the bandwidth to one third of the maximum capacity.

Throughput Efficiency

The introduction of IP gateways has breached the gap across disparate legacy networks to create connectivity in the tactical domain. However, using an IP core network to transmit legacy waveforms has low throughput efficiency and uncertain end-to-end Quality of Service (QoS). Both the Link 16 and EPLRS waveforms transmit for roughly 50% of the timeslot in order to compensate for jitter and for the distance between nodes. With the addition of FEC redundant bits the actual throughput of data can be as low as 6-20% of the transmission [5]. These waveforms were developed in the 1970s and 1980s and were designed for short messages and short payload voice. When these waveforms are encapsulated in an IP packet, which has its own significant overhead, the data throughput is reduced even further.

Assume that a non-IP waveform reaches an IP gateway with its data, header and synchronisation components. Denote the information component as I and the remainder of the transmission as $H1$, then the throughput efficiency for the message is $\eta = \frac{I}{I+H1}$. The non-IP message is encapsulated in an IP packet with IP, UDP and RTP headers, which can be grouped together as $H2$, which reduces the efficiency to $\eta = \frac{I}{I+H1+H2}$. As the IP packet traverses down the protocol stack it is split into MAC frames of size M, which produces $\lceil \frac{I+H1+H2}{M} \rceil$ frames. Each of these frames has its own MAC layer overhead, $H3$, making $\eta = \frac{I}{I+H1+H2+MH3}$. Then the DLL can add overhead of rate $R1$ for FEC and the RF layer can add overhead of $R2$ for R synchronisation. If we assume $R = R1 * R2$, then R can be relatively small, making

$$\eta = \frac{I}{I + H1 + H2 + MH3} \times R.$$

Clearly, throughput efficiency is a significant issue for IP-based tactical networks. Some research suggests that VoIP, which uses very small payloads, when traversing multiple tactical MANET hops, can have an effective throughput as low as 1% [5].

3.3.3 SINCGARS

Background

At the forward edge of the battle area forces continue to rely on single channel combat net radio (CNR) to coordinate force manoeuvre and operations. In such an environment it is radio voice communications that have primacy, but there is also a need for data networks in support of those force elements. The Single Channel Ground and Airborne Radio System (SINCGARS) waveform has been the mainstay of US front line tactical communications since it was introduced in the mid 1980s. SINCGARS is a CNR system that provides ground-to-ground and ground-to-air communications to on-the-move front line forces. SINCGARS, from its earliest days, was designed as a voice and non-IP data radio which could operate in a hostile electronic environment.

SINCGARS operates in the Very High Frequency (VHF) band from 30 to 87.975 MHz, with 2,320 25-kHz channels available, and uses frequency modulation (FM). Information can be transmitted and received in either a single-channel mode or a frequency-hopping mode depending upon the electronic countermeasures environment. The radio's modular design allows for commonality between lightweight manpack, vehicular, and airborne versions, with the heart of the system being a common receiver/transmitter (RT). During its lifetime there have been a series of improvement programs that have expanded and improved the SINCGARS data capability. Broadly, there are three generations of SINCGARS: the original ICOM radios, the System Improvement Program (SIP); and the Advanced System Improvement Program (ASIP). Nevertheless, the data capabilities of SINCGARS are much lower than EPLRS or the tactical trunk network, and in many ways exchanging data over CNR remains analogous to the *last mile* challenge faced by commercial telecommunication companies.

SINCGARS Operation

For voice communications SINCGARS uses analog FM, and for data communications it uses Binary Frequency Shift Keying (BFSK), with a frequency deviation of 6.5 kHz. During frequency hopping, symbols are transmitted on a channel randomly selected by a pseudo-noise sequence generator with the channel frequency hopping at a pre-determined rate of 111 times per second [2]. Given the transmission rate of 16 kbps, this makes SINCGARS a slow frequency hopping system.

Code	Effective Data Rate
RS(25, 18)/Parity	9.6 kbps
RS(25, 9)/Parity	4.8 kbps
RS(32, 12)/N(12, 5)	2.4 kbps
RS(32, 6)/N(12, 5)	1.2 kbps
RS(32,12)/Parity	5 kbps (Burst)

Table 3.2: FEC Codes for SINCGARS EDM

The older SINCGARS models employ repetition coding while the SIP radios use FEC, with Reed-Solomon codes, although they are backwards compatible. In order to accommodate the improved features and backward compatibility, two modes of operation are available for the user to select. The SINCGARS Data Mode (SDM) provides backward compatibility with the older model radios while the Enhanced Data Mode (EDM) can be selected to take advantage of the improved FEC. Data transmission rates range from 600 bps to 16 kbps, although information throughput is, at best, a much lower 4.8 kbps.

SINCGARS System Improvement Program (SIP)

Forward Error Correction. In 1996 the SINCGARS SIP launched a new generation of SINCGARS radios that provided a significant improvement in the data capability of the radio. Research had been conducted into FEC options for SINCGARS that would improve data throughput over a range of 20 - 30 km. The Enhanced Data Mode (EDM) implemented four FEC options, linked to the underlying data throughput. Each mode is a concatenated code, using two techniques in order to combat burst and random errors. The EDM FEC methods are listed in table 3.2, where N(12,5) refers to the Nadler code [8].

Introducing the EDM and implementing a DSP architecture provided an opportunity to improve throughput by changing how the SINCGARS waveform synchronised. Rather than relying on communication security for synchronisation, the EDM used the synchronisation properties of the frequency hopping algorithm, which significantly reduced the overhead in each transmission, as illustrated in figure 3.3.12.

Channel Access. The SIP radio also provided improved channel access by introducing a CSMA/CD algorithm to manage multiple access, as illustrated

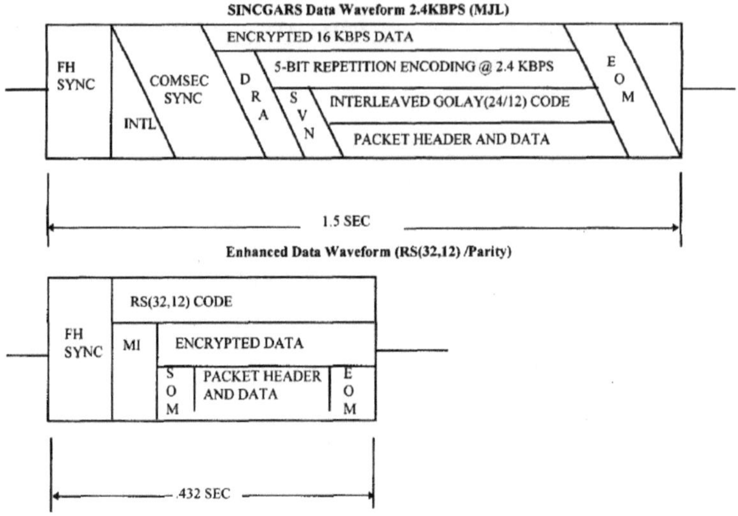

Figure 3.3.12: SINCGARS SIP Waveform Improvements (128 byte information packet) [8]

in figure 3.3.13. The original SINCGARS radio worked on an ARQ (request for retransmission) approach, which is inefficient in wireless networks. The SIP radios monitor transmission and after transmission ceases, all radios that wish to transmit are allocated a random slot from a frame of 32 possible slots. If no transmission has occurred when their slot time arrives then the radio starts transmitting. If a transmission is detected then the process commences again [8].

GPS. A GPS interface was included in the SIP. An external GPS could be connected to the radio and location data could then be included with each transmission. There was also a capacity for an internal GPS to be fitted.

Internet Controller Card (INC). The other major development in the 1996 program was the introduction of the Internet Controller Card (INC), which was urgently needed to support the emerging Tactical Internet. The INC introduced a packet routing capability both within (intranet) and between (internet) radio nets. The INC was housed in the vehicle amplifier adapter and a Mil-Std-188-220A data link layer protocol was used between the INC and the SIP radio. The INC mapped an IP address to each SIP radio on the network and could interface to other IP enabled networks, such as EPLRS. Data packet transmission was still managed by the radio, using the CSMA/CD protocol [8].

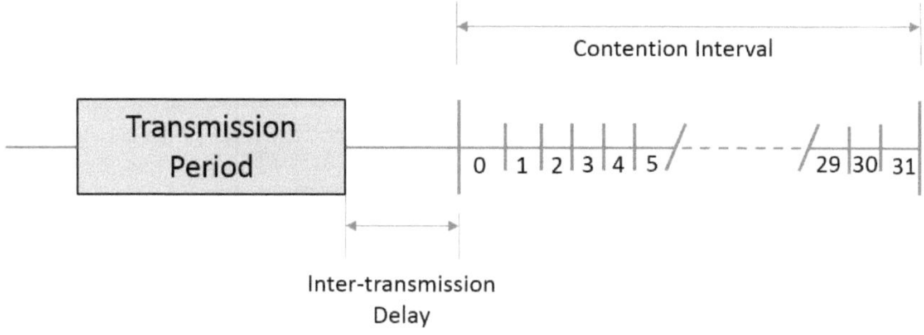

Figure 3.3.13: SINCGARS CSMA access protocol

The packet mode allows for the sharing of voice and data over the same operational net. A store and forward feature in the INC delays data while voice traffic is ongoing, and puts data on the net when the push-to-talk is released for voice. When the INC is loaded with initialization data, it will contain routing tables that identify the addresses of all members with which it is affiliated, as well as other radio nets that it can route to.

The host computers generate messages, along with the IP addresses of the individual(s) to whom the message is being sent. When a message reaches an INC, the INC looks up its routing table to determine whether that message is for a member of its net or whether it needs to be sent off to the next adjoining net. The packet mode will automatically continue this routing process until it reaches its destination.

The vehicle mounted INC is the predominant communications router for the tactical manoeuvre platforms participating in a SINCGARS enabled Tactical Internet. The INC routes data between SINCGARS and EPLRS, using commercial IP services to deliver unicast and multicast data packets that consist of command and control and situational awareness messages. Where EPLRS and SINCGARS gateways are collocated, such as at a tactical operations centre, the INC can use an Ethernet interface to the EPLRS RS. Two of the nineteen pins on the EPLRS connector are used as twisted pairs to provide for the 10Base-T Ethernet connection [2].

The protocol suite on the INC contains the functions typically found in an IP router, along with some special purpose modules for interoperability over tactical radio nets. The protocols are shown in figure 3.3.14 and essentially provide the bottom four layers of the OSI model. Special purpose EPLRS agents are included, which could be viewed as application layer functions [9].

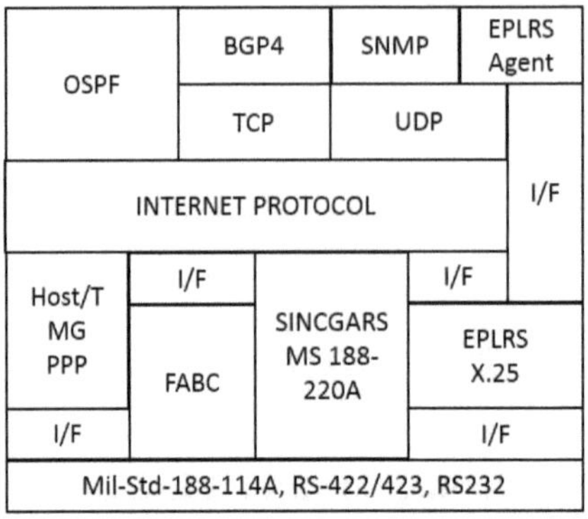

Figure 3.3.14: Internet Controller protocol architecture

At the physical layer, the INC supports Mil-Std-188-114, RS 422/423 and RS 232. The Advanced SIP version also includes an Ethernet interface. At the datalink layer the INC supports several protocols. The basic access protocol for a host computer is PPP, which allows almost any host to connect via an RS 232 interface. The SINCGARS radio communicates with the INC using Mil-Std 188-220, which allows voice and data sharing to be negotiated between the radio and the INC. Where the INC is used as a gateway into an EPLRS network, the Army Data Distribution System (ADDS) protocol suite is used. Since EPLRS is a non-IP network, the INC includes a particular EPLRS agent that modifies the message packets for compatibility with the EPLRS message format [9].

At the network level the standard TCP/IP protocol is supported, along with the necessary routing protocols. The SIP INC uses the Open Shortest Path First (version 2) (OSPF) routing algorithm, and for routing between autonomous networks it uses the Border Gateway Protocol 4 (BGP4). Network management is possible using the Simple Network Management Protocol (SNMP).

Advanced System Improvement Program (ASIP)

The early SIP model performed poorly in trials where there was a concentrated number of emitters, such as a major headquarters, where co-site interference significantly reduced the data throughput. To address these deficiencies, the Advanced SIP (ASIP) radio incorporates an enhanced system improvement program (ESIP) waveform. This waveform implements faster synchronization between radios, which reduces the interference between voice and data transmissions, thus reducing voice and data contention problems associated with shared voice and data networks [12]. The ESIP waveform includes optimisations to the algorithms of the noisy channel avoidance (NCA) scheme, the time of day (TOD) tracking scheme, and the end of message (EOM) scheme. The ASIP model is also half the size and one third of the weight of a SIP radio, mainly because of its digital signal processing architecture which reduces the number of required components. The VAA also received a more powerful 860 microprocessor; an Ethernet interface; enhanced protocols; and increased memory and buffer size [2].

References

[1] Department of the Army. Tb 11-5825-298-10-3 enhanced position location reporting system (eplrs) planner's manual. Technical report, US Army, 2009.

[2] US Army. *FM 6-02.53 Tactical Radio Operations*. Department of Defense, 2009.

[3] David Axe. Failure to communicate: Inside the army's doomed quest for the 'perfect' radio, January 2012.

[4] C. S. Bey. Airborne tactical data network gateways: Evaluating eplrs' ability to integrate with wireless meshed networks. Master's thesis, Naval Postgraduate School, Monterey, California, 2005.

[5] George F. Elmasry. *Tactical Wireless Communications and Networks: Design Concepts and Challenges*. Wiley, 2012.

[6] Jacob W. Fry. An analysis of mobile ad-hoc network performance to recommend a basis of issue for the u.s. army nett warrior system. Master's thesis, Naval Postgraduate School, Monterey, California., September 2010.

[7] L Hogan. G. Petch, P Straughair. Occasional paper: The impact of tactical communications on future military effectiveness. Technical report, Raytheon Australia Pty Ltd, 2012.

[8] Bradley J. Hamilton. Sincgars system improvement program (sip) specific radio improvements. In *Tactical Communications Conference*, pages 397–406. IEEE, 1996. 10.1109/TCC.1996.561109.

[9] L.J. Williams R. Schulman, R. Snyder. Sincgars internet controller -

heart of the digitized battlefield. In *Proceedings of the 1996 Tactical Communications Conference*, pages 417–421. IEEE, April 1996.

[10] Raytheon. Raytheon demos radio-based network in afghanistan, June 2013.

[11] M. Ryan and M. Frater. *Battlefield Communications Systems.* Argos Press, 2007.

[12] Paul Sass. Communications networks for the force xxi digitized battlefield. *Mobile Networks and Applications*, 4:139–155, 1999.

[13] Bernard Sklar. Reed-solomon codes, 2002.

[14] Bernard Sklar. *Digital Communications: Fundamentals and Applications.* Prentice-Hall, 1988.

[15] Deborah Tharp and Laird Wallace. Enhanced position location reporting system: Legacy system provides new technology for warfighters. *Space and Naval Warfare Systems Center San Diego Biennial Review*, 1:206–211, 2003.

4 Modern Tactical Data Architectures

4.1 Joint Tactical Radio System (JTRS)

IN the 1990s the increasing transfer of situational awareness data around the battlefield was confirming that tactical data networks were going to become the dominant communications channel for the majority of battlefield traffic. Voice command nets were still going to be the primary command mechanism for manoeuvre elements, but that command was increasingly supported by digital data from sensors, friendly position reporting systems and other battlefield management systems.

At the same time, there was a recognition in the US that there were too many different types of radio in the battlefield, which could sometimes hamper operations, rather than support them. A patrol could require one radio to communicate among themselves, another to reach back to their superior headquarters and yet another to communicate with passing air assets. In 1997, in a bold move, the US Defense Department established the Joint Tactical Radio System program. The program's aim was to develop a family of affordable, high-capacity tactical radios to provide both line-of-sight and beyond-line-of-sight C4I capabilities to the warfighters. This family of radios would cover an operating spectrum from 2 to 2000 MHz, and would be capable of transmitting voice, video and data. These software defined radios (SDR) were required to conform to open physical and software architectures. In this way, the program would offer small, single waveform, hand held radios through to multi-band, multi-waveform radios that could be re-programmed in the field for a range of missions. A fleet of 180,000 flexible radios was intended to replace the 750,000 radios in use by the US military [4].

The flagship radio was the Ground Mobile Radio (GMR), but after fourteen years, in October 2011, the GMR project was cancelled. The radio had produced poor results in field trials with the US Army, who claimed it was too big and too slow for use in tactical operations. The JTRS program was reorganised, creating the Joint Tactical Networking Center (JTNC) which has become a standards organisation for tactical software defined radios and their associated waveforms. Other aspects of the JTRS program were passed to different agencies within the US DoD. In the same period a number of vendors produced their own multi-band radios that are software based and can run some of the JTRS waveforms. JTNC now performs an assurance role by certifying these radios for use with the JTNC waveforms.

Despite the setbacks and cost of the JTRS program, it has contributed some significant milestones on the journey towards a robust, tactical IP-based MANET data architecture. The principal contributions have been defining the Software Communications Architecture (SCA) for tactical software defined radios, the Wideband Networking Waveform (WNW) and the Soldier Radio Waveform (SRW). At the tactical trunk level the JTRS program has also defined the High-Band Networking Waveform (HNW) and the Network Centric Waveform (NCW). At this stage, the future of tactical data networking appears to be married to these waveforms, although the timing for their adoption is unclear. Implementing the new waveforms requires new radios, which is a significant financial investment. At the same time, manufacturers of some of the existing radios have provided IP enhancements that make them attractive data networking solutions for the medium term.

4.1.1 JTRS Waveforms

The JTRS waveforms are designed to support the US vision of a Global Information Grid (GIG) which provides a seamless IP architecture across the battlefield. They will be implemented in software, providing a native IP layer, such that a software-defined GIG radio will be a complete IP node. Although the GIG is a US vision, the JTRS waveforms are likely to see service around the globe, since all major defence tactical communications companies have released, or are developing, JTRS compliant radios.

A waveform is the entire set of radio and/or communications functions that occur from the user input to the radio frequency output and vice versa. There are a range of JTRS waveforms, although they are no longer managed by a single US Defense agency:

- Tactical Edge (managed by JTNC [1])
    - Wideband Networking Waveform (WNW)
    - Soldier Radio Waveform (SRW)

  - Coalition Wideband Networking Waveform (COALWNW)
  - Mobile User Objective System (MUOS)
  - Legacy waveforms: SINCGARS, EPLRS, Link-16, High Frequency (HF), UHF SATCOM, VHF/UHF LOS, HAVEQUICK II, COBRA, JTRS Bowman
  - Joint Enterprise Network manager (JENM)
- Upper Echelon Tactical Waveforms
  - High-Band Networking Waveform (HNW)
  - Network Centric Waveform (NCW)

Each waveform is at a varying state of maturity or implementation, but vendors are seeking certification for some waveforms on their products. Most vendors are offering modern tactical communications equipment that complies with the Software Communications Architecture (SCA) and may be certified for one of the JTRS waveforms.

4.2 Wideband Networking Waveform (WNW)

The Wideband Networking Waveform (WNW) is designed to be the workhorse waveform at the tactical level. Conceptually it can replace EPLRS as the backbone network for the tactical GIG, just as the Soldier Radio Waveform (SRW) can be conceived as the replacement for the SINCGARS data network. The WNW delivers a guaranteed network throughput of 2 Mbps and uses the following bands of the frequency spectrum: 225-450 MHz, 1350-1390 MHz, and 1755-1850 MHz. In addition it has the following minimum transmission ranges: Air-to-Air (370 km), Air-to-Ground/Surface (370 km), Ground-to-Ground (10 km), Ship-to-Ship (28 km), and Ship-to-Shore (at least 28 km).

JTRS waveforms, such as WNW, implement the High Assurance Internet Protocol Encryptor (HAIPE) which supports multiple level security enclaves across the network. The HAIPE encryption layer uses an embedded processor for encryption in order to support parallel implementations of the HAIPE protocol, creating physically separate plain text IP enclaves. The plain text IP layer gives radio users access to an Ethernet port where VoIP terminals and IP COTS applications can be plugged in. The plain text IP layer is separated in hardware from the cipher text IP layer, and HAIPE insists that no topology or

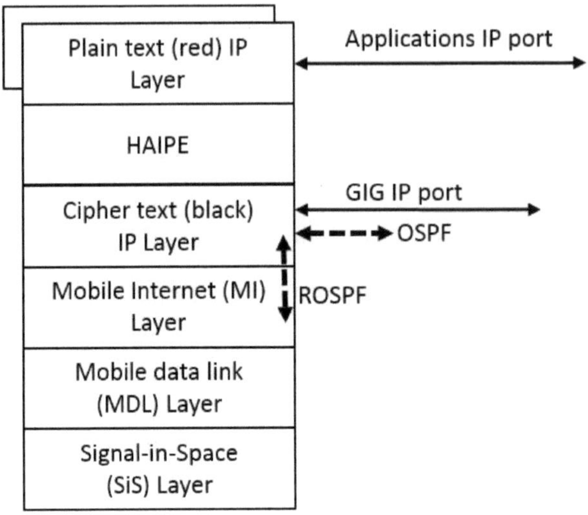

Figure 4.2.1: The WNW waveform protocol stack layers

routing information can cross from the plain text domain to the cipher domain [3]. The WNW protocol stack is illustrated in figure 4.2.1.

4.2.1 Cross-Layer Signalling

The standard OSI model is a protocol stack with independent functionality in each layer of the stack. As we have seen, the tactical wireless environment is much more constrained than the commercial environment, such that it is inefficient to insist on complete separation between the protocol layers. Instead, modern waveforms such as WNW use cross-layer signalling to introduce a measure of cooperation between the layers. For example, if the physical layer throughput is reduced due to signal fading, then it makes sense to advise the network layer to reduce the packet flow rate. The counter argument is that the layers are no longer modular and so it becomes difficult to purchase elements from different vendors and the implementations may become proprietary, rather than an open architecture. However, the benefits of cross-layer signalling in the wireless environment are such that it cannot be ignored and has been included in the modern waveforms. The key to interoperability is probably in defining the layer interfaces, rather than the internal mechanics of the signalling.

4.2.2 Cipher Text IP Layer

The cipher text IP layer in WNW uses the Open Shortest Path First (OSPF) routing protocol to maintain the network topology when connected to the GIG through a GIG IP port shown in figure 4.2.1. OSPF uses Link State Update (LSU) messages to check the routing state of neighbouring IP peers and to update known routes based on IP addressing. However, the network nodes connected to the radio stack are highly mobile, which can cause the link states to change regularly. Standard OSPF could consume all available bandwidth as the number of nodes increases on the network, so cross-layer signalling is used between the cipher text IP layer and the Mobile Internet (MI) layer so that these LSU are not discovered by the IP layer. Instead, a modified protocol is used called Radio OSPF (ROSPF) which allows for an increase in the number of nodes in a subnet and prevents over-flooding of the limited physical layer resources with LSU. In this way the cipher text IP layer can build an internal routing table over the radio WNW and an external routing table over the GIG port.

4.2.3 Mobile Internet (MI) Layer

The Mobile Internet (MI) layer provides what are sometimes called *MANET extensions* in the literature. The function of the MI layer is to reduce the burden on the IP layer due to the mobility of the network nodes attached through the radio stack. The MI layer maintains a multilevel link state routing topology of the WNW subnet, ensuring that the waveform maintains a connected topology that can be dynamically mapped to the routing table for IP delivery. The MI layer also controls data flow to prevent buffer overflow at lower levels and implements multilevel queuing to ensure QoS requirements.

The MI layer uses ROSPF to maintain link state information, with OSPF adjacencies provided by simulated *hello* packets containing link state information derived from the subnet routing [3].

4.2.4 Mobile Data Link (MDL) Layer

Architecture

Recall that with the EPLRS system slot allocation is established by the ENM when the network is established and is essentially static. This is an inflexible way to manage the scarce radio resources available to the network, so the WNW introduces a more dynamic capacity allocation algorithm for single and multi-hop broadcast called the *unifying slot assignment protocol* (USAP). USAP exchanges network management information with neighbouring nodes in order to assign the *optimum* number of slots to each neighbour and then coordinate these slots to avoid collisions [3]. The notion of *optimum* is derived by several heuristics that are tied to performance at the higher levels of the radio stack. A form of cross-layer signalling between the MI, MDL and Signal-in-Space (SiS) layers makes traffic demand from the upper layers and the current link condition from the lower layers, combine to influence slot allocation [2]. In this way, TDMA slots are concentrated in the nodes that are best positioned to relieve congestion. Cross-layer signalling from the SiS layer also reports changes in signal strength and symbol error rate, allowing the MDL layer to optimise the use of the physical layer resources. The architecture of the MDL layer is illustrated in figure 4.2.2. It consists of the following major functional modules [3]:

- Segmentation and reassembly,
- Queuing,
- Link adaptation,
- Orthogonal Domain Multiple Access (ODMA) channel access,
- Neighbour monitoring, and
- USAP resource management.

Segmentation/Reassembly and Queuing

The MDL layer breaks an IP packet into fragments to fill the TDMA transmission slots, and these fragments are then queued in the slots to await transmission. Fragment length is variable, depending upon the data rate, with more bytes fitting into a slot at a higher data rate. As the MDL layer pulls forward a fragment for transmission it sets the slot data rate. Where fragments are destined for multiple receivers the MDL layer uses the data rate that matches the slowest receiver.

At the receiver, fragments are reassembled into the original IP packet and then passed to the MI layer if the packet is a user data packet. If the received packet is an MDL control packet then it is passed to the internal neighbour monitoring or USAP resource management functions as shown in figure 4.2.2 [3]. The fragment header contains link transmission parameters that are extracted and

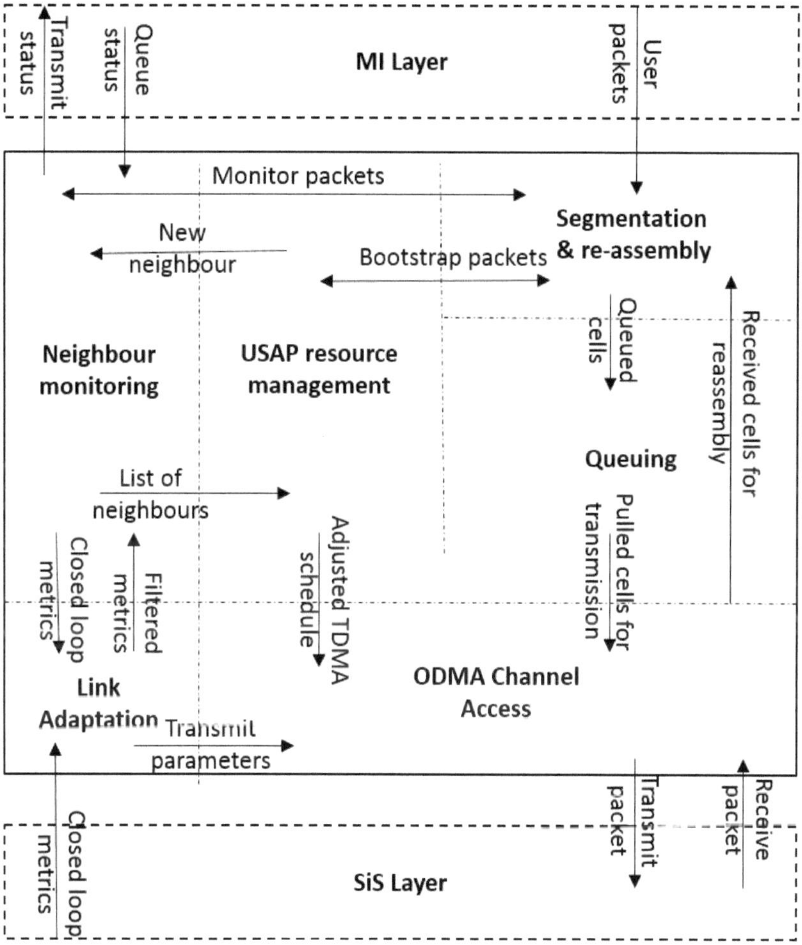

Figure 4.2.2: WNW MDL layer architecture

passed along with the raw SNR and symbol error rate (SER) metrics to the link adaptation function for processing.

Flow control is managed by cross-layer signalling between the MDL and MI layers. The MDL layer advises the state of the fragment queue and the MI layer advises what it has in the packet queue, including packets marked with a differentiated services code point (DSCP) indicating a higher preference, such as a 'call to fire' message. The MDL layer then adjusts its bid for additional TDMA slots to manage the flow of priority traffic.

Link Adaptation and the SiS Layer

The SiS layer can use several modes of OFDM and variable power settings, depending upon the condition of the link between the radio node and its neighbours. The modes are configured in a generic state machine which transitions from one state to another based on the received SNR and SER parameters. The incoming SNR and SER parameters are provided to the MDL Link Adaptation algorithm, which filters and blends them into the state machine control parameters. These are referred to as open loop parameters and are compiled for each incoming node pair, such as *node i* to *node j*, and *node k* to *node j*. These parameters are then reported back to the originating node in MDL control packets, becoming the closed loop feedback metrics. The MDL layer is biased to prefer the closed loop metrics in order to compensate for links that are different in each direction. In this way, the transmitting MDL layer can adjust its SiS layer to transmit in a configuration that matches the conditions at the receiver [3].

Unifying Slot Assignment Protocol (USAP)

The CSMA approach to sharing physical resources is best suited to local area networks which have a high level of physical resources compared to the traffic demand. Under these conditions collisions do not have a significant impact on data throughput because there is sufficient spare bandwidth to keep the collision rate relatively low. However, in tactical MANET the traffic demand is high compared to the available physical resources, so a more sophisticated sharing scheme is required. The Unified Slot Assignment Protocol (USAP) is a TDMA system that partitions the physical layer resources into two separate functions: control and data. The control segment is used for coordination messages between the nodes to avoid collisions.

A USAP frame is one second and within that frame there are $M \times F$ slots, where M is the number of timeslots in the frame and F is the number of frequency channels available. Each combination of time and frequency forms an MDL cell, with the collection of cells referred to as an orthogonal domain multiple access frame. There are three types of slots in the frame:

1. *Bootstrap Slots* are used to pass control information between nodes, such as USAP records. Each node in the network is assigned a bootstrap slot which it uses to inform its neighbours about slot reservations.
2. *Fixed reservation slots* support unicast data traffic. Once a slot is assigned to a link between *node i* and *node j,* no other node in a three hop radius uses this slot, ensuring that the transmission is contention free. At the same time, *node i* and *node j* cannot use the same timeslot on any other frequency.
3. *Rotating broadcast slots* support unicast, multicast and broadcast traffic. They act similarly to fixed slots but are shared in time as required by the traffic load.

In order to support the IP packet transfer, the allocation of slots is spread across each frame to try to present a reasonably consistent packet latency at the receiver. If slots are not spread then the delay in consecutive packet arrival can vary considerably, producing packet jitter [3].

Ensuring collision free operation is achieved by having a set of rules for slot assignment, based on each node knowing what its neighbours are doing over a three hop neighbourhood. Let $T(l)$ and $R(l)$ represent the transmitting and receiving nodes on each end of a link, l. To avoid collisions, $T(l)$ reservation is based on the following rules:

1. $T(l)$ cannot reserve timeslots that are already scheduled for incoming and outgoing transmissions for $T(l)$ and $R(l)$.
2. $T(l)$ cannot reserve slots used by incoming call transmissions to its neighbours.
3. $T(l)$ cannot reserve slots used by outgoing call transmissions from the neighbours of $R(l)$.

The effect of these rules is that spatial and frequency re-use can commence once the fourth hop has been reached [3]. The USAP protocol supports a connection-oriented framework, in which slot allocation is pre-configured along the entire link, or a connectionless framework, in which slot allocation is negotiated at each hop as the packet arrives.

Topology Control in Tactical MANET

With tactical MANET the underlying topology is constantly changing as nodes join and leave the network through manoeuvre or the effects of terrain. Therefore, the optimal route through the network from *node i* to *node j* will vary over time due to physical node changes as well as fluctuations in traffic density through particular nodes. The WNW uses a technique called the unifying connected dominating set (UCDS) algorithm to discover a minimum connected sub-graph through the network. The algorithm relies on distributed performance at each node, with nodes reporting their current state to their neighbours. The definition of the *optimal* path through the network will depend on several different parameters at each node, such as remaining battery life, average data rate, CPU power and available queue space. When physical changes occur, such as a node leaving the network, the UCDS algorithm has been shown to converge to a new dominating set within just a few cycles of the algorithm. More detailed information on the UCDS algorithm can be found in reference [3].

4.3 Soldier Radio Waveform

The Soldier Radio Waveform (SRW) is an IP based tactical MANET waveform which is intended to replace the legacy SINCGARS waveform. Like the WNW waveform, it provides an IP protocol stack in which the MAC, data link and physical layers have been modified to support tactical wireless operations. Unfortunately, there is little commonality between the modification in WNW and the modifications for SRW. The SRW protocol stack is shown in figure 4.3.1.

CNR Interface

The SRW has a dedicated CNR voice application that runs directly over the MAC layer and the data link layer, where CNR packets fit directly into MAC frames. This implementation has been designed to work only over the local subnet in support of platoon operations, although in theory a standard VoIP application could also be used. The waveform supports frequency channels in the spectrum range 225-420 MHz and 1350-2500 MHz.

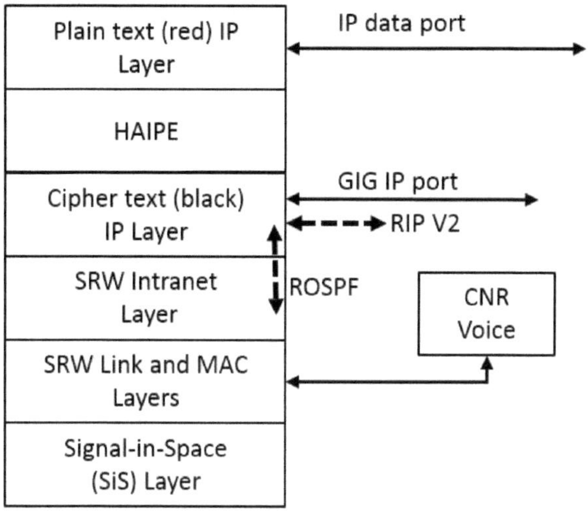

Figure 4.3.1: SRW protocol stack layers

Deployment

The waveform is designed for use on a broad range of platforms, including vehicles, dismounted soldiers, rotary wing aircraft, automated weapons, ground sensors and unmanned aerial vehicles. It is particularly intended for disadvantaged battlefield users, where space, weight and power are limited. In a man-pack or vehicle configuration the radio has a range of around ten kilometres and in a hand held configuration the range is around five kilometres [5].

Routing

The SRW uses the older Routing Information Protocol (RIP V2) to interface with autonomous networks, so it is not designed for backbone routing[1]. The SRW waveform can interoperate with the WNW by using the WNW for backbone routing. Internally, the SRW waveform uses the MANET designed ROSPF protocol. SRW supports data rates of 50 kbps to 2 Mbps, depending on the operational mode.

[1]RIP generates routing updates every 30 seconds, which becomes unwieldy on large networks.

4.4 High-Band Networking Waveform (HNW)

The High-Band Networking Waveform (HNW) is designed for upper echelon tactical deployment, operating as a ground-to-ground or ground-to-air waveform. Using an air node, it can form a wide area network with geographically dispersed ground nodes. HNW supports terrestrial and aerial line-of-sight communications, both on-the-move and at-the-halt and can provide high capacity, long range, full mesh networking for WIN-T backbone communications.

HNW provides mobile ad-hoc networking by automatically discovering and tracking moving nodes. It uses Directive Network Technology (DNT) to operate with a narrow beam, fast-switching, directional antenna for 360 degree plus aerial coverage. The automatic in-band neighbour discovery does not require operator intervention to establish links.

The HNW waveform can achieve a network throughput from 16 Mbps to 110 Mbps, depending upon modulation and FEC. It uses the C and Ku frequency bands to support ad-hoc ground-to-ground, ground-to-air and air-to-air communications. The range of the waveform is up to 45 km in a ground-to-ground, point-to-point, non-networking configuration, and 44 km in a ground-to air (UAV) configuration. It includes transmission security for user data and signalling.

The waveform can achieve spectral efficiency up to 2.2 bits/ Hz and frequency re-use is implemented through DNT and auto-scheduled timeslots. Demand assigned timeslots are also available, which allows sharing resources across the network and reducing the required spectrum. Similar to WNW, the HNW waveform dynamically adjusts throughput based on link conditions [5].

4.5 Software Communications Architecture (SCA)

One of the goals of the JTRS project was to develop a software defined radio (SDR) that could implement each of the modern waveforms easily, by re-programming the radio. Such a radio is capable of multiple mode operations, including variable signal formatting, data rates and bandwidths, within a single hardware configuration. The ideal SDR would have an analogue-to-digital converter at the antenna to transform the incoming electromagnetic wave into a stream of bits, with all further demodulation, error correction and other processing performed by software, as shown in figure 4.5.1.

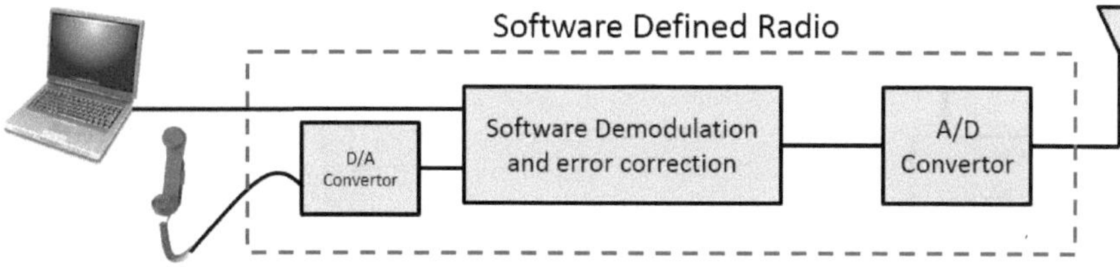

Figure 4.5.1: Software Defined Radio concept

This approach removes all of the special hardware usually associated with radios and replaces it with a set of software applications. Practically, such a scheme cannot be achieved because the conversion cannot be performed at radio frequencies at a sufficiently high rate, with the required quality. However, SDR schemes do exist that down convert the received signal to a suitable intermediate frequency and then convert the incoming signal to a digital data stream.

In practice, most modern tactical radios from large vendors, such as Harris and Raytheon, are software programmable radios (SPR) which use software extensively in the way that they implement the radio features. However, in order to ensure interoperability across vendors the JTRS project defined a Software Communications Architecture (SCA) for tactical radios. The goal was to provide a standard architecture for the deployment, management, interconnection and inter-communication of software components in embedded, distributed-computing communication systems [3]. Several vendors offer radios that are SCA compliant.

Achieving an SPR capability depends upon the ability to select and configure the appropriate hardware that supports the software required for a specific waveform. In this context hardware is not just the input or output devices, such as power amplifiers to set waveform power levels, but also the processing capabilities provided by general purpose processors (GPP), digital signal processors (DSP) or field programmable gate arrays (FPGA). The waveform interacts with the basic radio software, but that radio software must also provide to the waveform access to the underlying hardware. A common SCA ensures that software components deployed on a system are able to execute and communicate with the other hardware and software elements in the system. The ambition is to be able to consistently interoperate radios from multiple vendors, as illustrated in figure 4.5.2.

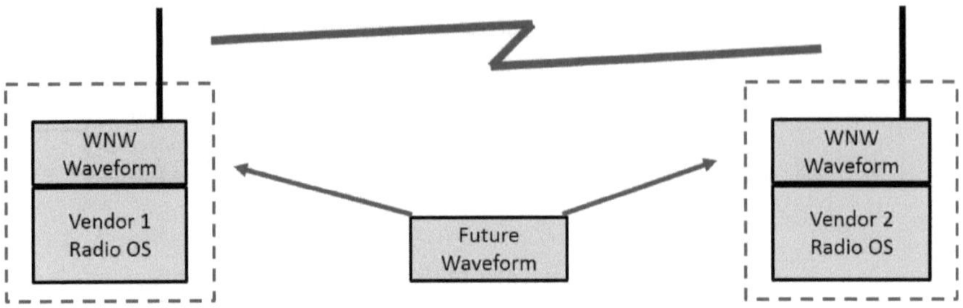

Figure 4.5.2: Vendor interoperability through the SCA

4.5.1 Core Framework

The SCA does not mandate particular hardware or software, but instead specifies the interfaces between the various hardware and software components in the radio. To do this, the SCA defines a set of abstract components for SCA compliant products. These components are:

- Base application components, which provide the management and control interfaces for all system software components.
- Base device components, which allow the management and control of hardware devices within the system, through their software interface.
- Framework control components, which control the instantiation, management and removal of software from the system.
- Framework services components, which provide additional support functions and services.

The SCA then defines the interfaces between these abstract components. The core framework interfaces are:

- Base application interfaces,
- Base device interfaces,
- framework control interfaces (FCI), and
- framework services interfaces (FSI).

The SCA layered approach is illustrated in figure 4.5.3. It is important to remember that the SCA standardises the component interfaces, but does not set any implementation requirements on how each vendor may implement the component. When a waveform is loaded it can access hardware devices by calling standard functions offered by the base device components, which then interface to the hardware through a base device interface. The waveform

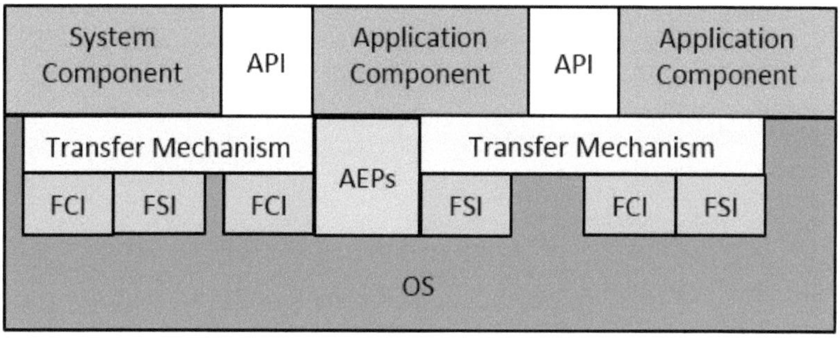

Figure 4.5.3: SCA layer diagram

can access standard supporting functionality through the framework services components, which offer standardised functionality.

Through the use of abstraction the SCA core framework supports waveform interoperability between multiple vendors who can offer radios at varying states of software programmability or software definition. This provides a pathway towards a true software defined radio and ultimately towards cognitive radio.

References

[1] Joint Tactical Networking Center. Our products, Accessed 17 Mar 2014.

[2] George F. Elmasry. The progress of tactical radio from legacy systems to cognitive radios. *IEEE Communications Magazine*, 51:50–56, October 2013.

[3] George F. Elmasry. *Tactical Wireless Communications and Networks: Design Concepts and Challenges*. Wiley, 2012.

[4] GlobalSecurity.org. Joint tactical radio system: Programmable, modular communications, Accessed 14 March 2014.

[5] Tracy Mann. Jtrs/win-t: networking waveform quick reference sheets. *Army Communicator*, 33(3):72–76, June 2008.

5 Data Beyond-Line-of-Sight

ALL major military nations agree that future warfare will require dispersed, autonomous, mobile forces that can draw on or receive large amounts of situational awareness and sensor data [3]. Dispersal will cause formations to operate beyond-line-of-sight (BLOS) of their tactical gateway parent node, yet require reliable, high capacity data links. Traditionally, such links were provided by the trunk communications system which would establish a chain of microwave radio relay terminals between the gateway node and the mobile formation. However, radio relay repeater assets are usually truck mounted and take time to deploy and set up because they raise hydraulic mast antennas in order to cover the long distances between relay nodes. In an era when a brigade headquarters might prop for six hours, this was an acceptable solution, but in today's battlefield a faster solution is needed.

Satellite communications are considered as one option for pushing high capacity data into BLOS situations. However, satellites have their own limitations and there is a significant difference between the data rates that can be provided at-the-halt (ATH) and on-the-move (OTM). Data capacity is closely tied to power and bandwidth, but also to the physical size of the satellite antenna. Mobility dictates a small satellite dish, since it needs to be mounted on a vehicle, but small dishes struggle to achieve high data rates. As with all engineering, providing satellite data communications to mobile military forces is a trade off of one characteristic against another.

Tactical satellite terminals tend to be power limited, and end-users are typically receiving more information than they are transmitting. Therefore, in this chapter we will focus on understanding the satellite downlink and exploring how the mobile tactical domain affects the communications performance of the satellite system. A more comprehensive description of satellite communications can be found in reference [6].

	Frequencies	Used for..	Typical Systems
UHF	243-318 MHz	Tactical voice and data (up to 64 kbps) Penetrates disadvantaged environments.	MUOS, UFO, SKYNET
L band	800-2150 MHz	Tactical voice and data (~500 kbps)	BGAN
C Band	6/4 GHz	Commercial TV Telecommunications (~60 Mbps)	INTELSAT
X Band	8/7 GHz	High bandwidth data (~500 Mbps)	WGS, SKYNET
Ku Band	14/12 GHz	High bandwidth data (~1 Gbps)	OPTUS, PALAPA
Ka Band	31/18 GHz	High bandwidth data (~2-3 Gbps)	WGS

Table 5.1: Common satellite frequencies for military communications

5.1 Satellite Fundamentals

5.1.1 Frequencies

The common satellite frequencies used in military operations are listed in Table 5.1.

5.1.2 Downlink Power

Satellite downlink power is usually given as the Effective Isotropic Radiated Power (eirp) and is measured in Watts, or equivalently in dBW (EIRP) when decibel notation is used. The eirp is a way of expressing the *effective* power radiated towards the earth and comprises two separate effects: the power provided to the transponder by a High Power Amplifier (HPA) on the satellite, and the directivity of the satellite antenna which focusses that power towards a segment of the earth's surface.

For example, the recent UK SKYNET 5 satellite provides fifteen transponders, with a 160 Watts HPA per transponder. Commercial satellites may use a HPA at perhaps 40-60 Watts. Now if that power were radiated using an isotropic radiator on the satellite then it would radiate in all directions equally, in the shape of a sphere. Clearly, that would be a waste of power, since radiating into deep space doesn't communicate with anyone, but it turns out that the *concept* of an isotropic radiator is a useful benchmark to use when we compare the effective power radiated by the satellite towards earth. We use the *gain* of an antenna as a measure of its directivity, or its ability to focus the radiated power onto a restricted portion of that imaginary sphere, rather like the difference between a torch and a candle.

Consider a lossless isotropic radiator that emits P_t Watts. As the signal travels R metres the original P_t Watts are distributed on the surface of a sphere with radius R. The flux density, ϕ, measures the power density in Watts per square metre (Wm^{-2}) at any point on the surface of the sphere. Hence, at a distance R, the flux density is given by $\phi = \frac{P_t}{4\pi R^2}$.

In reality, there is no such think as a perfect isotropic radiator. Instead, each antenna has a certain gain that amplifies the transmitted power in a particular set of directions. The more directive an antenna is, the higher its gain. We include the transmitter antenna gain, G_t, in the equation as $\phi = \frac{P_t G_t}{4\pi R^2}$ and the term $P_t G_t$ is called the eirp because it represents the magnitude that an isotropic radiator would need to have in order to produce the same power intensity as this antenna does in its direction of maximum gain. Hence, $\phi = \frac{eirp}{4\pi R^2}$.

5.1.3 Power Received at the Earth Terminal

The earth terminal sees the flux density from the transmitter and collects the signal carrier power through the receive antenna's effective aperture, A_e. The effective aperture is measured in square metres such that the received carrier power can be expressed as $C = \phi A_e$ (Watts). The effective aperture is composed of two parts: the first is the average collecting area, which is the same for all antennae at a particular frequency. The average collecting area does not depend on the shape or geometry of the antenna, only the wavelength, λ, of operation, and is calculated as $A_{ave} = \frac{\lambda^2}{4\pi}$[1]. For each antenna, the average collecting area is then modified by the geometry of the antenna, because the geometry changes the directivity of the receive antenna. The parameter used to measure directivity is the antenna gain, G_r. Hence, the effective area of an antenna, A_e,is given by $A_e = \frac{\lambda^2}{4\pi} G_r$ and

$$\begin{aligned} C &= \frac{P_t G_t}{4\pi R^2} \cdot \frac{\lambda^2 G_r}{4\pi} \\ &= \frac{P_t G_t G_r}{L_p} \end{aligned} \tag{5.1.1}$$

where $L_p = \left(\frac{4\pi R}{\lambda}\right)^2$ is called the *path loss* and C represents the signal power available at the input to the receiver/demodulator.

Note: The inclusion of λ in the expression for the path loss is for convenience only, as it allows us to separate out the receive antenna gain. The λ is actually related to the average collection area of the receive antenna. The amount of attenuation of electromagnetic waves through space is not dependent upon the frequency of the wave. Nevertheless, in the literature comments are often made that imply that if a system shifts to a higher frequency then it will incur a greater path loss. This is not really true; the *loss* occurs because the collection area of the receive antenna becomes smaller as the frequency increases.

5.1.4 Earth Terminal Performance

Geosynchronous satellites orbit the earth at a height of 36,000 km above the equator, so when a signal finally reaches earth the signal power is very low. Therefore, the satellite earth terminal needs high sensitivity in order to distinguish the satellite signal from general electromagnetic noise. One of the significant noise sources at satellite frequencies is the electronic thermal noise in the earth station itself due to the random fluctuations of electric currents. It is well known that this *noise power* is represented by $P_n = kTB$, where B is the bandwidth over which the measurement is made, T is the temperature in Kelvin and k is Boltzmann's constant. Other noise sources include atmospheric noise and cosmic noise which are often represented using an *antenna noise temperature*. The total noise power is therefore described as

$$N = kT_{sys}B \tag{5.1.2}$$

[1] See Appendix 1 for a derivation of the average collecting area.

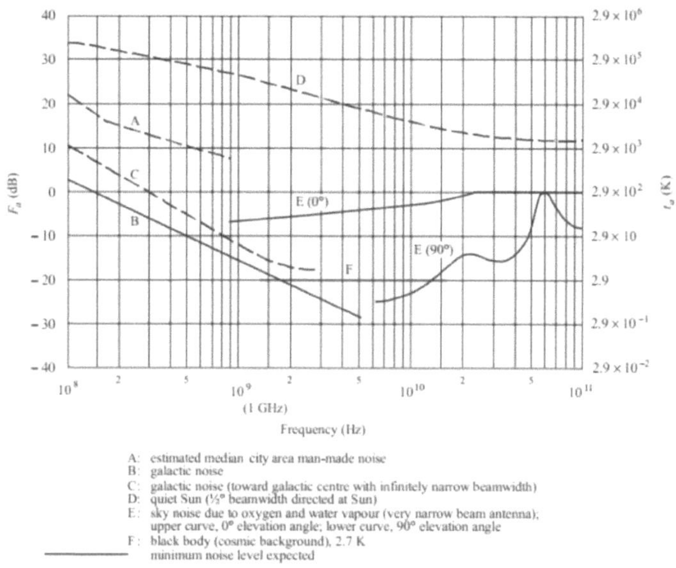

Figure 5.1.1: Noise source contributions [ITU-R P.372-11]

where T_{sys} is defined as the overall receive system temperature, which represents the noise power generated in the receiver system from all sources. Figure 5.1.1 shows the typical noise contribution form a range of sources.

The overall earth terminal performance is therefore governed by two competing parameters: the *gain* of the receive antenna, and the *system noise temperature* of the receiver. Earth terminal performance is usually quoted in satellite literature using the parameter $\frac{G}{T}$. A larger $\frac{G}{T}$ implies a better performance. The combination of the carrier power received from the satellite and the noise introduced through the antenna, receiver and demodulation process at the earth terminal, as represented by the $\frac{C}{T}$, provide the key link budget parameter, the carrier-to-noise ratio, or $\frac{C}{N}$.

5.1.5 Signal to Noise Ratio

In satellite systems the capacity on a particular link, such as the data rate that can be achieved, or the minimum acceptable error performance, depends upon the baseband ratio of signal power to the noise power, or the $\frac{S}{N}$, at the output of the demodulator. A distinction is made in satellite literature between the carrier-to-noise ratio, measured at the input to the receiver, and the signal-to-

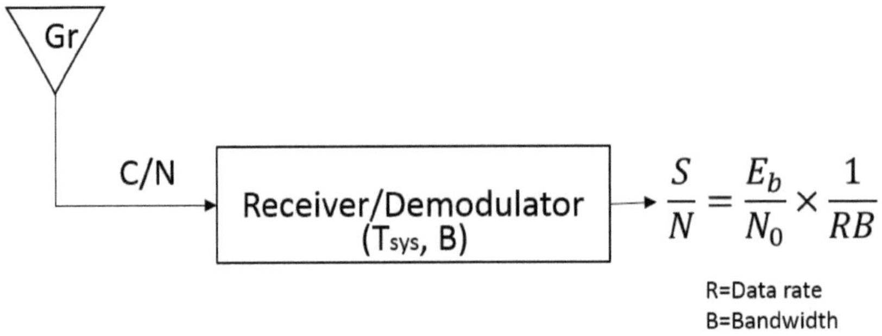

Figure 5.1.2: $\frac{C}{N}$ vs $\frac{S}{N}$.

noise ratio, $\frac{S}{N}$, measured after the demodulation process. This distinction is illustrated in figure 5.1.2.

Using equations 5.1.1 and 5.1.2, the signal-to-noise ratio is given by

$$\begin{aligned}\frac{S}{N} &= \frac{P_t G_t G_r}{L_p} \times \frac{1}{kT_{sys}B} \\ &= \frac{eirp}{L_p}\left(\frac{G_r}{T_{sys}}\right)\frac{1}{kB}\end{aligned}$$

or, independent of bandwidth,

$$\frac{S}{N_0} = \frac{eirp}{L_p}\left(\frac{G_r}{T_{sys}}\right)\frac{1}{k}. \tag{5.1.3}$$

This form is useful from an engineering perspective as it separates out the influence of satellite transmission, path loss and the earth station performance.

For digital satellite signals an alternate equation for the signal-to-noise ratio can be constructed by recognising that the demodulated signal power is equal to the demodulated energy per bit multiplied by the demodulated bit rate, or

$$\frac{S}{N_0} = \frac{E_b \times R}{N_0} \tag{5.1.4}$$

where $E_b$is the energy per bit in the demodulated signal and R is the demodulated data rate. By combining equations 5.1.3 and 5.1.4 we arrive at the useful link budget equation that measures the data rate achievable over a particular satellite system

$$\frac{eirp}{L_p}\left(\frac{G_r}{T_{sys}}\right)\frac{1}{k}=\frac{E_b\times R}{N_0}.$$

This equation is usually expressed in decibel form as either [6]

$$\begin{aligned} R &= EIRP+\frac{G}{T}-L_p+228.6-\frac{E_b}{N_0} \\ EIRP &= R-\frac{G}{T}+L_p-228.6+\frac{E_b}{N_0} \\ L_m &= EIRP+\frac{G}{T}-L_p+228.6-R-\frac{E_b}{N_0} \end{aligned} \tag{5.1.5}$$

where L_m is the link margin allocated to the link.

5.1.6 Antenna Gain

We have already seen that the gain available to an antenna depends on frequency, through the average collecting area, and the geometry of the antenna as expressed in $G=\frac{4\pi}{\lambda^2}A_e$. For a parabolic dish antenna, typically used in satellite communications, the effective area is the circular area of the dish opening, $\frac{\pi D^2}{4}$, where D is the diameter of the dish. Therefore, the theoretical gain of a satellite dish is given by G=$\left(\frac{\pi D}{\lambda}\right)^2$, or practically by

$$G=\eta\left(\frac{\pi D}{\lambda}\right)^2 \tag{5.1.6}$$

where η is the efficiency coefficient.

Equation 5.1.6 offers some interesting insights into satellite link performance. For a dish of fixed size, as the frequency increases so does the antenna gain. Since gain measures directivity, what this means is that at higher frequencies the antenna is more directive, or the beam-width becomes narrower, focussing the energy into a smaller cone. It is tempting to conclude that higher data rates can be achieved by simply working at higher frequencies, but this is an illusion. The path loss definition, L_p, also has a $\frac{1}{\lambda^2}$ factor and in the link budget equation 5.1.5 it can be seen that any increase in antenna gain through increasing frequency is offset by an increase in the path loss.

Again considering equation 5.1.6, if frequency is held constant then gain is improved if the dish size increases. This a genuine gain because as the dish

System	Frequency	Channels	Transponder Power	EIRP
MUOS Legacy	UHF UHF	17x25 kHz 21x5 kHz	- -	26 dBW Global beam 20 dBW
MUOS WCDMA (Chip rate = 3.84Mcps)	UHF	16 beams 4x5 MHz carriers per beam 512 channel codes per carrier	620W (28 dBW)	51 dBW (single beam) 43 dBW (all beams)
Skynet 5	X-band	8 x 20 MHz 3 x 22 MHz 4 x 40 MHz	160W (22dBW)	50.5 dBW (spot beam peak) 40 dBW (global beam peak)

Table 5.2: Example satellite downlink parameters [4]

aperture increases the antenna is able to "catch" more of the signal energy, improving the signal-to-noise ratio.

5.2 Tactical Satellite Examples

To illustrate the use of satellite communications for tactical data we will use two model satellites in the UHF band and in X band. These are commonly used frequency bands for military operations. The UHF specifications are based on the US Mobile User Objective System (MUOS), which also includes a legacy UHF payload; the X-band specifications are based on the UK Skynet 5 satellite. Table 5.2 lists some of the published satellite downlink parameters for each system.

The baseband modulation scheme will be QPSK and we will consider several FEC schemes incorporating Reed-Solomon and convolutional encoding that are commonly implemented on terminals. For a $P_b = 10^{-6}$ QPSK requires an $\frac{E_b}{N_0} = 10.6$ dB. Table 5.3 lists the additional coding gain achieved using

Reed-Solomon Code Rate	Conv Code rate	Coding Gain
-	1/2	4.9 dB
-	3/4	3.9 dB
-	7/8	2.8 dB
R(256, 239)	1/2	7.4 dB
R(256, 239)	3/4	6.6 dB
R(256, 239)	7/8	5.2 dB

Table 5.3: Coding gain for QPSK at $P_b = 10^{-6}$, using several coding schemes

TES Parameter	Value
Frequency	7.5 GHz
Slant distance (Assumes location at 65°latitude.)	39,837 km
G/T	22.7 dB/K
Path Loss	202 dB

Table 5.4: Transportable earth station parameters

several variants of FEC. The first three schemes use convolutional coding only and the final three schemes are concatenated FEC, with a convolutional inner code and an R(256, 239) Reed-Solomon outer code.

5.2.1 Formation Headquarters with Large Antenna

Formation headquarters are the data hubs for the tactical domain, containing high capacity rear links to strategic networks, and managing the trunk communications resources to push data forward to lower level users. Traditionally, the divisional headquarters has held this role, but increasingly the need for data is pushing high capacity solutions towards brigade headquarters and forward units. High capacity military satellite links require a large satellite antenna, which is either towed on a trailer or fitted to a purpose-built vehicle solution. A typical large tactical satellite dish might be 2.4 metres in diameter and multi-band terminals can operate in the X-band, Ku-band and Ka-band frequency ranges. Table 5.4 lists the key down link parameters for a typical X-band transportable earth station.

Assume that we are using QPSK (2 bps/Hz), which requires an $\frac{E_b}{N_0}$= 10.6 dB at a $P_b = 10^{-6}$, with a rate 1/2 FEC coding scheme that introduces a coding

SPITFIRE Parameter	Value
Frequency	254 MHz
Slant Distance	39,837 km
Tsys	1028 K
Gr	6 dBi
G/T	-24.1 dB/K
Path Loss, Lp	173 dB

Table 5.5: SPITFIRE technical parameters

gain of 4.9 dB. The Skynet 5 satellite has eight 20 MHz X-band transponders, with a peak EIRP of 40 dBW into the global beam. To calculate the data rate that could be supported into this terminal we use the data rate formula from 5.1.5, assume a link margin of 6 dB and include the coding gain.

$$\begin{aligned} R &= EIRP + \frac{G}{T} - L_p - L_m + 228.6 - \frac{E_b}{N_0} + G_{FEC} \\ &= 40 + 22.7 - 202 - 6 + 228.6 - 10.6 + 4.9 \\ &= 77.6\,\text{dBbps} \\ &= 57.5\,\text{Mbps} \end{aligned}$$

Therefore, this transportable X-band terminal can support a trunk bit rate of 57.5 Mbps. Since rate 1/2 encoding is being used, the throughput is 29 Mbps.

5.2.2 UHF At-The-Halt (ATH)

For vehicle mounted operations, or dismounted operations, UHF satellite equipment has been used to provide reliable BLOS communications in a variety of terrains. In the past there has been a short set up time required to orient the antenna towards the satellite, so these systems have not been available on-the-move. However, more recent systems have included satellite tracking technology and a vehicle mounted radome to maintain communications whilst the vehicle is moving.

The Raytheon SKYFIRE is a tactical UHF transceiver, developed from its widely popular predecessor, the SPITFIRE. Table 5.5 lists the key down link parameters of the SPITFIRE.

Consider a mobile user who halts his vehicle in order to download an update to the current situational awareness picture. The user is operating a SPITFIRE radio and has interfaced to the tactical internet using a 25 kHz channel on the legacy UFO package on the MUOS system, with a link margin of 10 dB to compensate for fading and other losses.

Using uncoded QPSK and the data rate version of the the link budget formula 5.1.5, we have:

$$\begin{aligned} R &= EIRP + \frac{G}{T} - L_p - L_m + 228.6 - \frac{E_b}{N_0} \\ &= 26 + (-24.1) - 173 - 10 + 228.6 - 10.6 \\ &= 36.9\,\text{dBbps} = 4.9\,\text{kbps} \end{aligned}$$

However, the SPITFIRE uses a convolutional coding FEC scheme with a coding rate of 1/2. From table 5.3 we can see that this introduces a coding gain of 4.9 dB, which significantly improves the achievable rate to

$$\begin{aligned} R &= 36.9 + 4.9 \\ &= 41.8\,\text{dBbps} = 15.1\,\text{kbps}. \end{aligned}$$

In practice, devices like this run at pre-fixed data rates, such as 9.6 kbps, which are linked to an appropriate FEC scheme.

5.2.3 UHF On-The-Move (OTM)

In order to achieve an on-the-move capability the UHF satellite antenna must be either omni-directional or include a tracking solution to maintain directional integrity towards the satellite. Although satellite tracking introduces more complexity, it usually realises an improved antenna gain, since the antenna main beam can remain focussed towards the satellite.

Consider an on-the-move solution with a tracking antenna housed in a vehicle radome with a gain of 5 dBi. The user is communicating through the WCDMA MUOS system, which can support a data rate of 384 kbps. If we assume a rate 1/2 FEC scheme and QPSK modulation then the system will provide a spread spectrum coding gain of 10 dB, and a FEC coding gain of 4.9 dB. The system noise temperature for a mobile phone device is around T_{sys} =1500 K and hence the $\frac{G}{T} = -26.7$ dB/K. Therefore,

$$\begin{aligned} L_m &= EIRP + \frac{G}{T} - L_p + G_{SS} + G_{FEC} + 228.6 - R - \frac{E_b}{N_0} \\ &= 43 - 26.7 - 173 + 10 + 4.9 + 228.6 - 55.8 - 10.6 \\ &= 20.4\,\text{dB} \end{aligned}$$

which represents a very healthy link margin. The principal reasons that MUOS can achieve such high rates are the increased power available in the transponder and the ability for each user to use the entire transponder bandwidth as required.

5.2.4 Blue Force Tracker

The US Blue Force Tracker (BFT) system reports location data to a common operating picture, which is then re-broadcast into the area of operations. BFT was originally deployed at the beginning of the Iraq war and has been improving ever since. An early complaint was the latency of updates, which made the system unusable during battle, since the commander could not be sure that the system was reporting the correct current location of his forces. Version two of the system claimed to reduce the latency from minutes to seconds. BFT runs over a commercial L-Band satellite network provided by ViaSat, although version three is likely to run over the US Wideband Global Satellite (WGS) system.

Although a Battlefield Management System screen is required to view the position of friendly forces, a BFT reporting module has also been developed for dismounted patrols. The unit periodically (several times per minute) transmits its location to the satellite and that location data is added to the common operating picture. The BFT module is a small, low power device with no satellite tracking capability, yet it successfully broadcasts regular location updates. It is worthwhile considering how this can be achieved.

Terminal

The specifications of the terminal are likely to be similar to those provided here:

- Power ~ 5 W
- Antenna Gain ~ 1.5 dB

- Hence, EIRP ~ 8.45 dBW

Estimated Uplink Parameters

Parameter	Value
EIRP (Hand terminal to satellite)	8.45 dBW
Uplink Frequency	1.5 GHz
Slant Distance	39,837 km
Path Loss	188 dB
Satellite G/t (Looking toward Earth)	-1.2 dB/K
$\frac{E_b}{N_0}$	10.6 dB
P_b	10^{-6}
G_{FEC}	4.9 dB
Link Margin	20 dB

A substantial link margin would be required to compensate for the erratic movement of troops on the ground, as well as the canopy variation, or urban reflections, while moving. Applying these values to the link equation leads to the following sustainable bit rate.

$$\begin{aligned} R &= EIRP + \frac{G}{T} - L_p + 228.6 - \frac{E_b}{N_0} + G_{FEC} - L_m \\ &= 8.45 + (-1.2) - 188 + 228.6 - 10.6 + 4.9 - 20 \\ &= 22.15 \text{ dBbps} \\ &= 164 \text{ bps.} \end{aligned}$$

Using these parameters the terminal can only support a bit rate of 164 bps, which is inadequate for regular position reporting. A transmission of around 3 kbps is required, which would become 6 kbps once FEC is applied. BFT uses a form of DSSS to introduce sufficient processing gain to sustain the higher bit rates. The bandwidth of channels at L-Band can vary between 3 MHz and 70 MHz, so there is sufficient spectrum available.

To achieve $R = 6$ kbps we need a processing gain of $G_{SS} = \frac{6000}{164} = 36.6 =$ 15.6 dB. Therefore, the chipping code needs to be $R_c = 36.6 \times 6000 = 220$ kbps, and using QPSK this transmission could be sent in a channel of around 200 kHz.

5.2.5 Network Centric Waveform (NCW)

The JTRS program has defined a satellite waveform for use in the Super High Frequency (SHF) range which is designed to support a full mesh IP network over the satellite, using a TDMA architecture. The NCW waveform uses a dynamic scheduler to allocate TDMA slots as required, based on traffic demand and traffic priority. Traffic priority is configurable, which allows this waveform to support a quality of service regime.

Implementing IP over satellites in a TDMA architecture requires some modification to the protocol stack to cope with variable propagation delay between the satellite and the terminal, which can be as long as 120 ms [2]. Therefore, in a hub-spoke network, a two-hop delay of 240 ms must be expected. On top of this, there is a queuing delay introduced by the TDMA architecture which can create a total delay of nearly one second.

5.3 Bandwidth or Power Limited

Broadly, the data rate limitation on the downlink of tactical satellite channels can be categorised as either bandwidth limited or power limited, depending upon the terminal type. Large earth stations tend to be bandwidth limited, where the limitation is in the bandwidth of the transponder on the satellite. Power limitations are less of an issue, since the size of the antenna dish can collect a relatively large amount of the signal. By contrast, small tactical terminals that have small antennae tend to be power limited, where the limitation is on the power available to the satellite transponder. The small dish can only collect over a small area, so to support high data rates the initial power levels need to be relatively high.

5.3.1 Bandwidth Limited Channel Example

Consider a military SHF satellite with a 40 MHz transponder and an EIRP = 35 dBW in an earth coverage beam. If we assume a spectral efficiency of 1 bps/Hz then the transponder can support a maximum bit rate of 40 Mbps.

Now consider a large fixed military earth terminal with a dish diameter of eighteen metres and a $\frac{G}{T} = 39$ dB/K. If the modulation scheme requires an $\frac{E_b}{N_0} = 8.6$ dB and we need a 3 dB link margin, then the satellite EIRP required to support this channel is:

Parameter	Value
Data Rate 40 Mbps	76 dBbps
Path Loss 7500 MHz	202 dB
Terminal G/T	39 dB/k
Boltzmann's Constant	-228.6 dB
$\frac{E_b}{N_0}$	8.6 dB
Link Margin	3 dB
Satellite EIRP	**22 dBW (160 Watts)**

Under this scheme the terminal uses all of the available transponder bandwidth, but only 5% of the transponder power.

5.3.2 Power Limited Channel Example

Using the same military satellite, consider a tactical earth terminal with a small antenna and a resulting $\frac{G}{T} = 0.1\text{dB/K}$. If the terminal can draw all of the satellite transponder power then the achievable data rate is:

Parameter	Value
EIRP (earth coverage)	35 dBW
Path Loss 7500 MHz	202 dB
Terminal G/T	0.1 dB/k
Boltzmann's Constant	-228.6 dB
$\frac{E_b}{N_0}$	8.6 dB
Link Margin	3 dB
Data Rate	**50.1 dBbps (102 kbps)**

Under this scheme the terminal uses all of the transponder power but only 0.25% of the available bandwidth.

5.3.3 Shannon Interpretation of Limiting

Claude Shannon's famous 1948 paper [5] on information theory demonstrated that the maximum rate at which error-free communication can be transmitted on a channel is given by

$$H = B\log_2\left(1 + \frac{C}{N}\right) \tag{5.3.1}$$

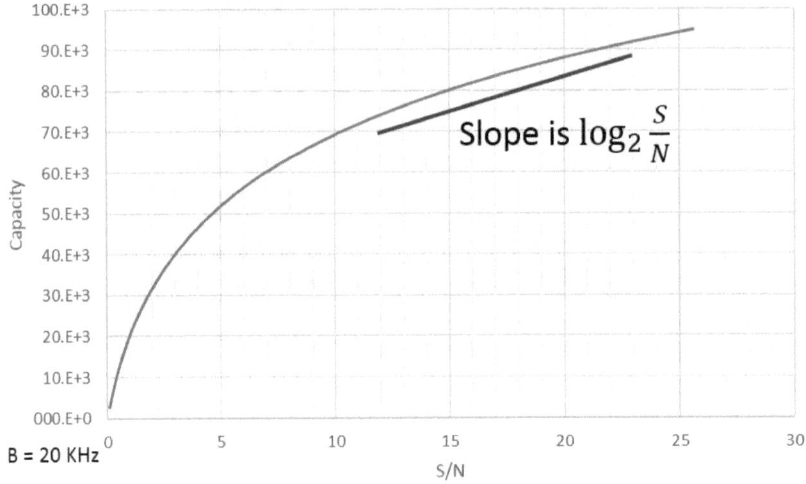

Figure 5.3.1: Bandwidth limited channel

where H is the rate of information transmission, or the channel capacity, B is the channel bandwidth and $\frac{C}{N}$ is the carrier power to noise power ratio. This result provides an interesting view on the problem of channel limitation and the relationship between power and bandwidth.

Bandwidth Limited

Consider again the large terminal example where received power is more than adequate. Equation 5.3.1 indicates that if $\frac{C}{N} \gg 1$ then $H \propto B$

$$H \simeq B \log_2 \left(\frac{C}{N} \right)$$

and any increase in bandwidth will provide a commensurate increase in channel capacity. However, $H \propto \log_2 \left(\frac{C}{N} \right)$ so increasing the power will only modestly increase the channel capacity. So if power is not a problem, then bandwidth becomes the limiting factor, as shown in figure 5.3.1.

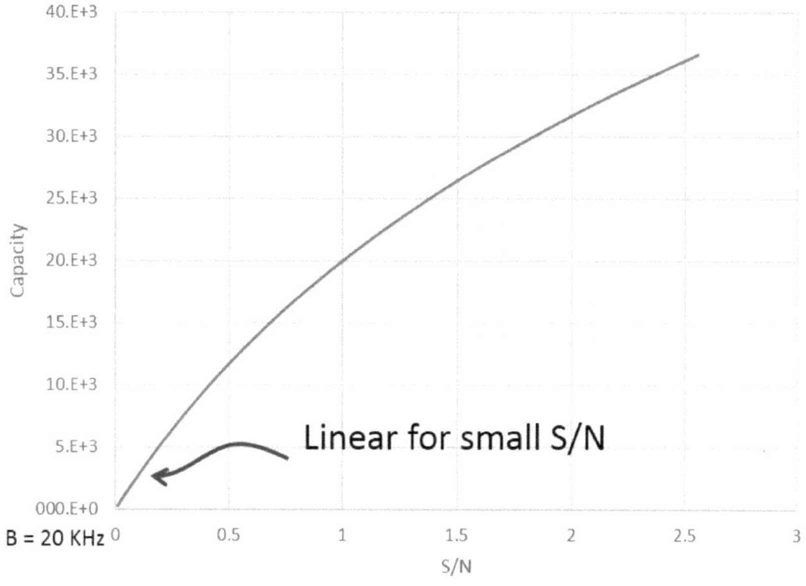

Figure 5.3.2: Power limited channel

Power Limited

Consider again the small terminal example where received power is much lower, or $\frac{C}{N} \ll 1$. In this case equation 5.3.1 indicates that

$$\begin{aligned} \lim_{B\to\infty} H &= \lim_{B\to\infty} B \log_2 e \log_e \left(1 + \frac{C}{N}\right) \\ &= \lim_{B\to\infty} \frac{C}{N_0} \log_2 e \log_e \left(1 + \frac{C}{N}\right)^{\frac{N_0 B}{C}} \\ &= \frac{C}{N_0} \log_2 e \end{aligned} \tag{5.3.2}$$

from the definition of e, the base of natural logarithms[2]. Hence, if bandwidth is much greater than the information rate then the channel capacity becomes directly proportional to the signal power, independent of bandwidth, as illustrated in figure 5.3.2.

[2]See Appendix 2 for the derivation of this expression.

Fundamental Limit

The transmitted bit rate can be represented by the relationship $C = E_b R$, where E_b is the energy per bit, R is the transmitted bit rate and C is the carrier power. Shannon's equation describes the relationship between power, bandwidth and the maximum possible information rate that can be achieved with an arbitrarily small error performance, but offers no method of achieving that information rate. However, if we assume that we have developed the optimum coding scheme such that the bit rate equals the information rate, or $R = H$, then from equation 5.3.2, $H = \frac{E_b H}{N_0} \log_2 e$ or

$$\frac{E_b}{N_0} = -1.6 \text{ dB}.$$

This is the famous Shannon limit, illustrated in figure **5.3.3**, which shows that an optimum coding scheme would require an $\frac{E_b}{N_0}$ of -1.6 dB to achieve an arbitrarily small probability of bit error. Compare this with the earlier commercial examples that use coding schemes that require around 4 to 10 dB for successful operation. However, modern schemes are approaching the Shannon limit.

5.4 High Frequency Waveforms

Although satellites are widely used across the battlefield, the transponder bandwidth in the space segment is a finite resource which is quickly consumed in an intense conflict. Other options exist for supporting tactical BLOS data, including high frequency (HF) radio. HF transmissions achieve long range by refracting the radio waves through the ionosphere, and back to earth at the distant location, as shown in figure 5.4.1.

Because the charge density within the ionosphere varies during the day and night, managing HF communications requires regular monitoring of the circuit quality and regular frequency adjustments to match the ionospheric behaviour. This extra work created a reputation for HF that it was a difficult medium to work in, which was true when compared to simple VHF and UHF systems. The extent of refraction depends on the frequencies used, but also on the take-off angle of the electromagnetic wave. Take-off angle can vary with antenna type, as illustrated in figure 5.4.2, but it can also change with the

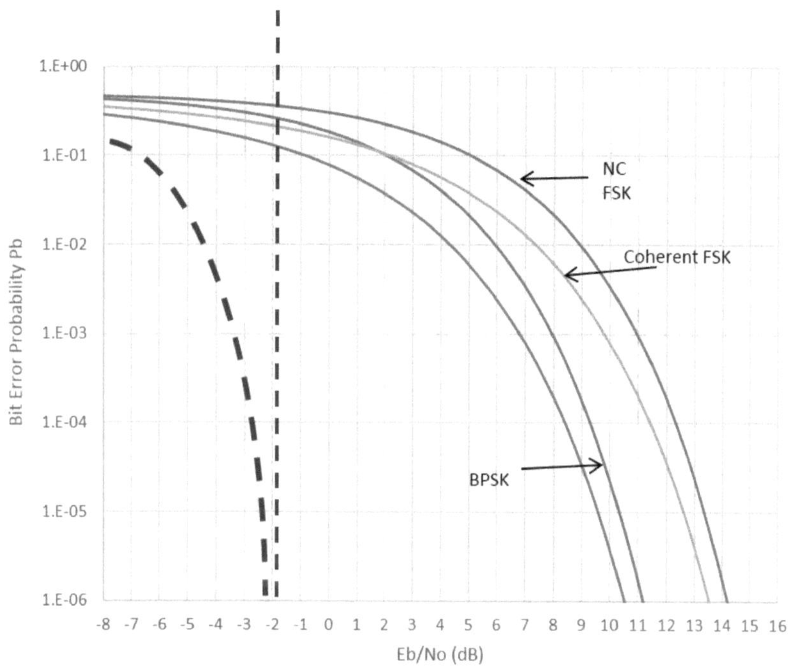

Figure 5.3.3: Fundamental Shannon limit

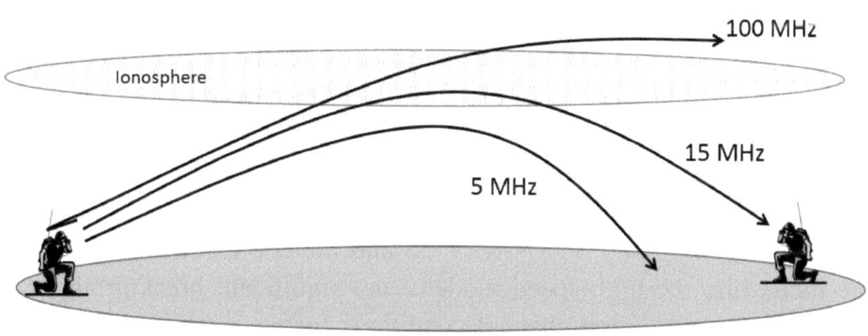

Figure 5.4.1: HF propagation

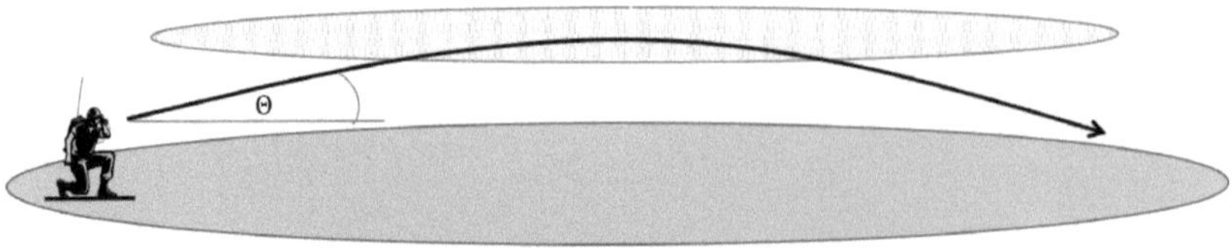

(a) Monopoles have a low take-off angle

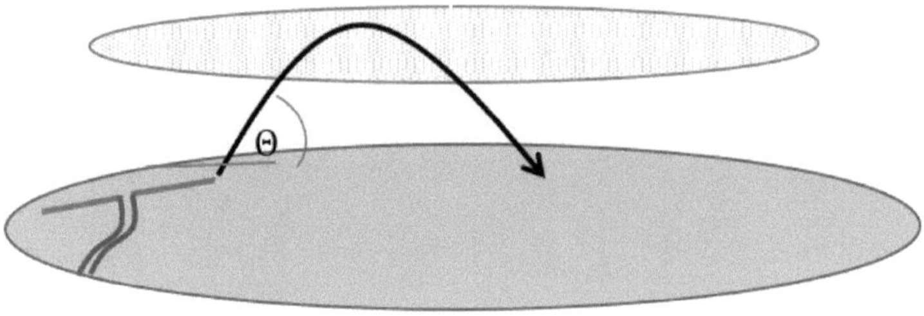

(b) Take-off angle varies with antenna height

Figure 5.4.2: Variation in take-off angle

height of an antenna above the ground plane. Achieving and maintaining a tactical HF link required constant attention.

HF fell out of favour in the 1970s, with the advent of satellite communications, but it has had a resurgence in recent years. Significant development in functions like Automatic Link Establishment (ALE) has created HF systems that manage themselves, relieving the communications operator of much of the management burden. By automatically selecting the best available frequency, the capacity of a HF radio can be maintained. Of course, the bandwidth available in the HF spectrum is limited, but if no other BLOS system is available then HF is a viable alternative for low-rate data communications.

The bandwidth of a HF channel is typically 3 kHz and, assuming a 1 bps/Hz spectral efficiency, the data rate over HF is typically around 2400 bps. A new generation of higher speed waveforms was developed in the late 1990s to provide reliable communications from 3600 to 9600 bps in a single 3 kHz channel. A two-channel version of the modem can provide rates as high as 19,200 bps. These waveforms are documented in MIL-STD-188-110B, and STANAG 5066 and STANAG 4538 and include extensive use of FEC. Note that while the high-speed modems are significant breakthroughs over HF, they do require good channel conditions for acceptable bit error rates. Consequently, the current generation of data systems dynamically adapt the data rate to the available channel [1].

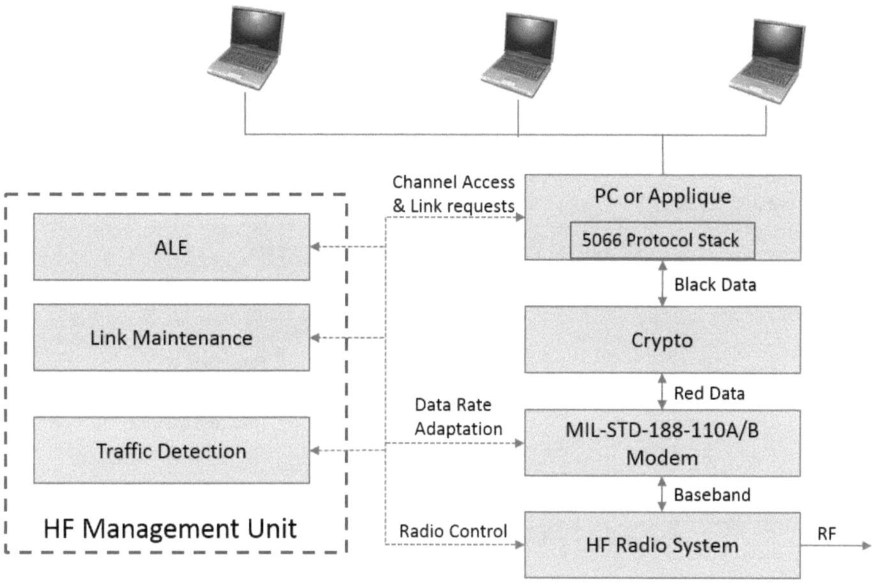

Figure 5.4.3: STANAG 5066 Architecture [1]

STANAG 5066

STANAG 5066 has become the de facto choice for many military organizations because it provides a cost effective capability by using existing radios and modem technology. It uses second generation technology and many interoperable implementations can be obtained from a range of vendors.

STANAG 4538

STANAG 4538 provides improved performance through faster linking, lower latency and improved network capacity. It provides a performance improvement of two and a half to five times better than STANAG 5066, particularly at low signal to noise ratios. STANAG 4538 uses third generation HF technology.

References

[1] Harris Corporation. *Radio communications in the digital age*, volume 1. Harris Corporation, 2 edition, 2005.

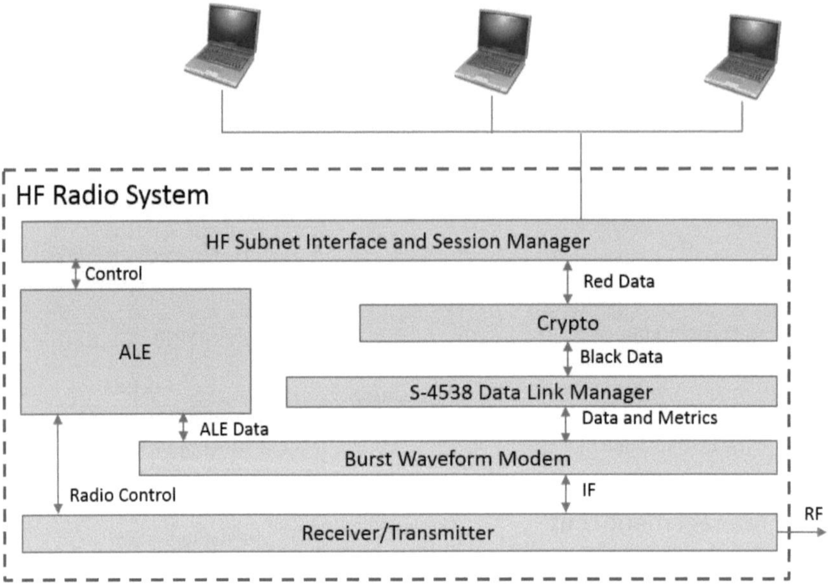

Figure 5.4.4: STANAG 4538 Architecture [1]

[2] George F. Elmasry. *Tactical Wireless Communications and Networks: Design Concepts and Challenges*. Wiley, 2012.

[3] L Hogan. G. Petch, P Straughair. Occasional paper: The impact of tactical communications on future military effectiveness. Technical report, Raytheon Australia Pty Ltd, 2012.

[4] C.K. Matassa. Comparing the capabilities and performance of the ultra high freqency follow-on system with the mobile user objective system. Master's thesis, Naval Postgraduate School, Monterey, California, June 2011.

[5] Claude E. Shannon. A mathematical theory of communication. *Bell System Technical Journal*, 27:379–423, 623–656, 1948.

[6] H.G. Suyderhoud W. Pritchard and R.A. Nelson. *Satellite Communications Systems Engineering*. Prentice Hall, 1993.

6 Data Below Brigade

6.1 Introduction

IN chapter 3 and chapter 4 we discussed some of the technical issues around the radio waveforms used for data networks below brigade level. In particular, we considered the SINCGARS radio as an example of a legacy data radio and the Soldier Radio Waveform as an example of a modern IP-based waveform developed specifically for tactical data networking. These types of networks are still referred to as combat net radio (CNR) to differentiate them from the higher capacity radios found further up the tactical chain. Combat net radios are typically bandwidth limited, so in this chapter we will consider the topology of CNR data networks and the services that need to run over these networks.

6.2 Voice Data Contention

Forward edge tactical commanders make heavy use of voice radio nets, particularly once operations are underway. For this reason the command voice net still remains the first communications priority for these units, although data now plays an essential role in maintaining situational awareness. Several schemes have been used to manage voice and data contention, depending upon the resources available, each with their advantages and disadvantages.

Separate Voice and Data Networks

One way to ensure that the voice network is always available is to run the data network on a separate channel. Given that CNR is traditionally a single channel radio this effectively means using two different radios: one for voice and one for the data network. This arrangement is illustrated in figure 6.2.1.

This approach was used by several nations in their first forays into the realm of digital battle management because in the late 1990s CNR did not offer suf-

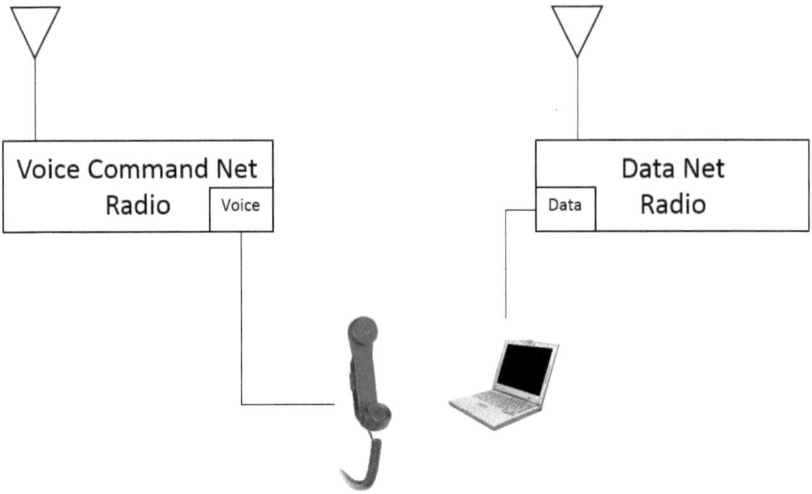

Figure 6.2.1: Separate voice and data networks

ficient functionality to operate voice and data reliably through the same radio. Most armies were running a command net and an administration net, so it was straightforward to convert the administration net into a data network. The advantage of this approach is that there is never any contention between the voice needs and data needs of the end users, ensuring that the commander has voice access as needed. The disadvantage is the requirement for two radios instead of one, which is a significant logistics burden. The additional radio requires space and consumes some of the weight and power profile of a vehicle. Due to the burst nature of data it could also create a network of radios that have a low activity rate for much of the time.

Single Channel with Voice Primacy

A single channel radio can operate on a joint voice and data network if additional functionality is included to sense the request for the voice transmission and to then suspend and buffer the data packets until the transmission is over. Some CNR come with this functionality included in the radio, but others will require a third party module to sense and manage the voice transmission requests. We have already encountered the SINCGARS SIP radio that operates in this way, using the INC to interface with a local LAN. The early Australian Battlefield Command Support System (BCSS) used a similar approach to interface with the legacy Raven radio. Figure 6.2.2 illustrates how the INC

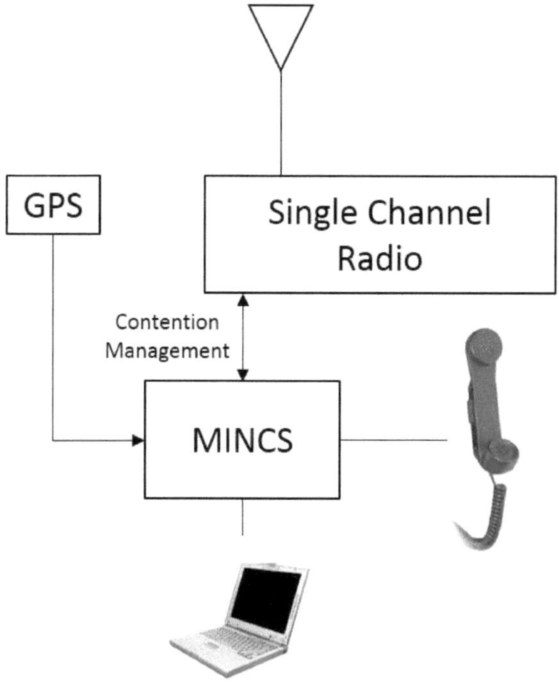

Figure 6.2.2: Single channel radio with integrated data

handshakes with the radio to negotiate access for the data transmission in between voice transmissions.

Note that the voice transmission can continue to be an analogue transmission that uses the existing radio circuitry, or it can be a digital voice transmission.

Pseudo-Second Channel

A third way to manage voice and data contention is to digitise the voice transmission and then multiplex the voice and the data into what effectively becomes two digital channels transmitted across a single analogue channel. This approach makes the entire transmission a digital transmission with voice and data being broadcast simultaneously. Of course, the voice will consume some of the data bandwidth, reducing the underlying data rate, but with effective multiplexing algorithms this effect can be minimised. This concept is used in the SINCGARS ASIP radios and in several modern radios and is illustrated in figure 6.2.3.

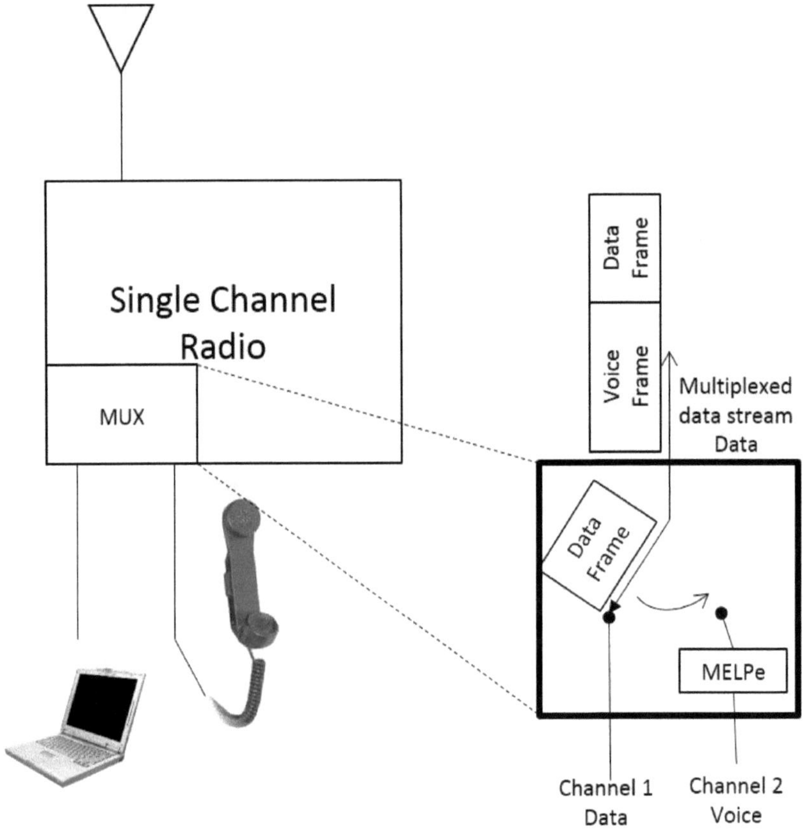

Figure 6.2.3: Multiplexed voice and data

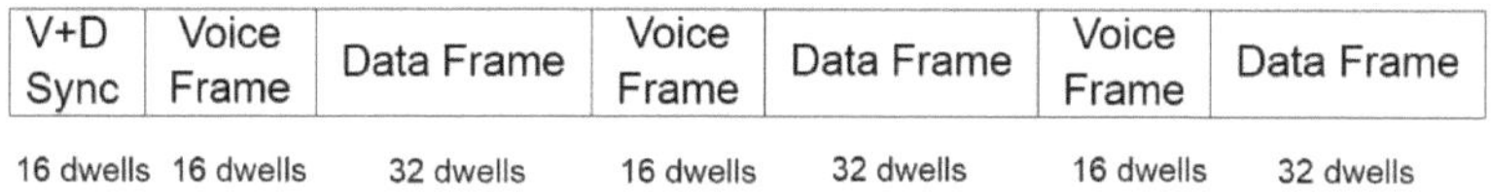

Figure 6.2.4: SINCGARS Voice and Data Frame

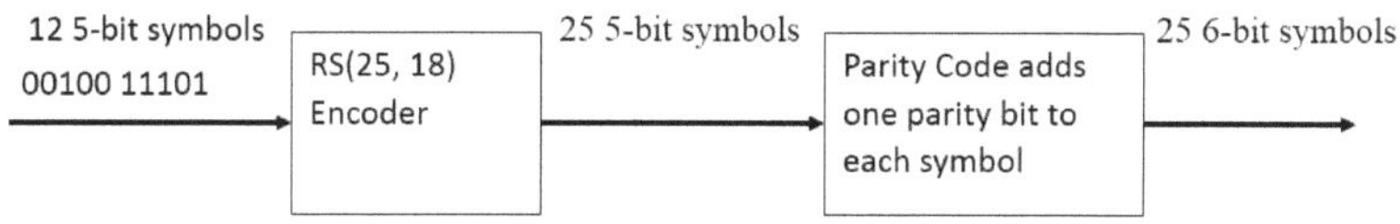

Figure 6.2.5: SINCGARS coding scheme

Example: SINCGARS Voice and Data Packet

In the SINCGARS ASIP program the tension between voice and data was addressed by creating a Voice and Data (V+D) frame that could be used when the radio was in frequency hopping data mode. The frame structure is shown in figure 6.2.4 [3].

The frame commences with a frequency hopping synchronisation pattern for 16 hop dwells [2]. The remainder of the transmission alternates between a voice frame for 16 hop dwells, followed by a data frame for 32 hop dwells. Consider a data transmission at a rate of 9.6 kbps, which therefore uses the Reed-Solomon(25,18) encoding option in the SINCGARS radio, with a parity outer code. Reed-Solomon is a non-binary code, with each symbol being 5 bits in the SINCGARS. The coding process is illustrated in figure 6.2.5. The Reed-Solomon encoder takes 18 input symbols, with each symbol being a 5 bit data word (90 bits), and produces a 25 symbol codeword (125 bits). The codeword is fed into the parity encoder which takes each symbol (5 bits) and adds a parity bit to form the final symbol (6 bits). This process effectively produces a Reed-Solomon (25, 18) codeword with 6 bit symbols.

For this transmission each data frame comprises 24 codewords. The codewords are block interleaved across 25 hops with each hop containing one symbol from each codeword, as shown in figure 6.2.6.

When in packet data mode, voice is digitised at 2.4 kbps using the Mixed-

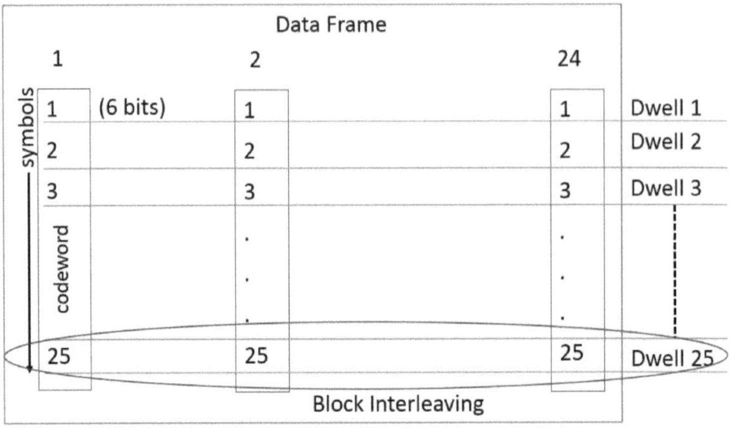

Figure 6.2.6: SINCGARS interleaving

Excitation Linear Prediction (MELP) standard, and within the SINCGARS radio it is then treated as a 2.4 kbps data stream, with the error coding embedded within the MELP algorithm. In orer to maintain continuity of speech, the 2.4 kbps voice packet must be able to be packed into the 16 dwell voice frame for transmission, so the size of the voice packet is dictated by the size of the super frame.

The SINCGARS hops at 111 hops per second, or 9 ms per dwell, resulting in a voice-plus-data super-frame of 432 ms. However, at the start and end of each data frame the radios in the net may need to switch from transmit mode to receive mode, which can take up to 35 ms, producing an actual data transmission time of around 362 ms. Of that time, 288 ms are used by the data frame, leaving around 74 ms for the voice frame. Since the output of the radio is 16 kbps, the voice frame can transmit around 1,184 bits. The MELP encoder produces a 432 ms voice data packet of size $2400 \times .432 = 1036$ bits, which can be transmitted in the voice frame.[1]

6.3 Digital Voice Signals

Nearly all modern tactical radios that are designed for use below brigade offer a digital voice channel. Those radios that emphasise wideband data network-

[1] Some of these figures are approximate, based on limited publicly available data.

ing usually offer Voice Over IP (VoIP) as the voice system, and treat the voice channel as simply another data packet, albeit with a higher priority than other data sources. Achieving this requires a modified TCP/IP protocol in order to introduce a quality of service (QoS) field to differentiate IP packets carrying voice signals. The more traditional approach to digitised voice in tactical radios is to use a synchronous voice coding scheme such as Linear Predictive Coding (LPC) or Continuously Variable Slope Delta Modulation (CVSD).

Linear Predictive Coding

Speech can be modelled as a collection of formants, sibilants and plosive sounds which can be characterised by a set of numbers. LPC is a narrowband voice coding algorithm that characterises speech by estimating the formants, removing their effects from the speech signal, and estimating the intensity and frequency of the remaining buzz. The process of removing the formants is called inverse filtering, and the remaining signal after the subtraction of the filtered modelled signal is called the residue.

The numbers which describe the intensity and frequency of the buzz, the formants, and the residue signal, can be transmitted over a relatively narrow bandwidth channel, which makes this approach to digitising voice popular in HF radios. LPC synthesizes the speech signal by reversing the process, using the buzz parameters and the residue to re-create the source signal, and then using the formants to create a filter. Passing the speech signal through the filter re-creates the speech sounds. Since speech signals vary with time, this process is done on short chunks of the speech signal, which are called frames and generally 30 to 50 frames per second give intelligible speech with good compression. LPC signals can be transmitted as a 2.4 kbps signal through a 3 kHz HF channel, often producing voice sounds that are understandable, but lacking recognition quality.

Mixed Excitation Linear Prediction (MELP/MELPe)

In the 1990s an improved low bit rate vocie encoder was introduced called Mixed Excitation Linear Prediction (MELP). MELP was enhanced over the following ten years through research and an open competition. The current version is called MELPe and it has been defined in MIL-STD-3005 and STANAG 4591. MELP provides a better quality voice signal than LPC and has become the standard narrowband vocoder.

Continuously Variable Slope Delta Modulation (CVSD)

CVSD is a modulation scheme that starts with a voice reference level known by the transmitter and the receiver and has an agreed step size by which the transmitter and receiver will vary the signal. If the transmitted voice signal increases in voltage then the CVSD modulator transmits a digital one to indicate that the receiver should also step up by one voltage level. If the transmitted voice signal decreases in voltage then the CVSD modulator transmits a digital zero to indicate that the receiver should also step down by one voltage level. Ideally, the transmitter and receiver will mirror the effects of the user as they speak.

In reality, a speech signal may undergo a rapid rise in voltage such that the rigid stepping process cannot keep up with the speech signal. To overcome this "variable slope" problem the CVSD algorithm increases the step size if several digital ones are transmitted in a row. Hence the rate at which the receiver increases voltage will grow if the source signal voltage increases over several successive bits. Once a change in direction is sensed, the step size reverts back to the default value and the process commences again.

CVSD encodes at one bit per sample, so that audio sampled at 16 thousand samples per second is encoded at 16 kbps. VHF tactical radios routinely offer 16 kbps CVSD modulation for digitised voice signals.

6.4 Battlefield Management Systems

6.4.1 Introduction

Battlefield Management Systems (BMS) have become the key application running on data networks in the forward battle areas. At the core of any BMS is the ability to exchange information between comrades and also up and down various echelons of command. This can be achieved using trunk radio or satellite communications across long distances, or via tactical radios for communications between ground units and individual soldiers. Principally these communications are used to carry voice, data and imagery traffic. There is no single definition of what a BMS is, but across nations they generally offer the following services:

1. **Common Tactical Picture**. The BMS is designed to provide each node with a common view of the disposition of friendly, adversary and neutral forces on the battlefield. When the picture changes, those changes

are promulgated across the network, keeping everyone up to date with the latest view.

2. **Blue Force Tracking**. This is the ability to display on an electronic battle map the location of all friendly forces in the immediate area of operations. Typically, this service is linked to the Global Positioning System (GPS) and automatically transmits an update of the current location of each friendly force element. The BMS receives this update and plots the location on a digital battle map in use by the local commander.
3. **Mapping and Terrain Analysis**. Many BMS offer a digital mapping product to support terrain analysis and planning. Blue force tracking is integrated with the mapping product.
4. **Operational Messaging**. The BMS provides an efficient method to exchange operational information between nodes on the battlefield using data and imagery. The data can include written orders, situation reports, intelligence summaries and requests for assistance, while the imagery can include everything from still photographs taken of a suspected enemy position by a thermal imaging camera, to real-time video footage provided by an Unmanned Aerial Vehicle (UAV) orbiting overhead. Many BMS also include a chat system that allows operators to make short comments or ask short questions of operators at other nodes.
5. **Operational Planning Tools**. A BMS can provide speciality tools to support particular military elements. Typical examples include artillery planning and calculations, intelligence support tools and combat engineering calculators. The aim is to integrate such tools with the core BMS architecture in order to produce a single Common Tactical Picture that can be viewed by all nodes operating the BMS.

The key communication principle underpinning the BMS is a low bandwidth requirement, since the systems usually find greatest use in forward areas of the battlefield, which typically have the smallest communication pipes. To that end, the systems usually make use of pre-defined messages that require only a small number of bits per message.

6.4.2 Example Systems

United States

In the late 1990s the BMS concepts first became reality as the US Army developed its plans for "Force XXI", which was to define the shape of the army for the new century. These plans revolved around developing a lighter, more flexible and easily deployed force with greater strategic mobility, that would rely more on its information superiority to achieve success than its combat weight. These ideas were refined in a series of experiments and trials which incorporated the first group of systems which would provide a degree of automation of the command and control process. These included the Manoeuvre Control System (MCS), the Advanced Field Artillery Tactical Data System (AFATDS), the All Source Analysis System (ASAS), the Tactical Airspace Integration System (TAIS) and Force XXI Battle Command Brigade and Below (FBCB2), which with others collectively formed what was known as the Army Battle Command System (ABCS).

FBCB2 is the tactical system that supports commanders below brigade level and can be deployed down to squad level. FBCB2 is connected over line-of-sight radio networks, principally EPLRS but sometimes SINCGARS, and consists of a ruggedised terminal that can be vehicle mounted. Amongst other things, the system provides a terrain analysis tool, the location of friendly forces and expected adversary locations, which combine to tell the commander when enemy fire is likely to be encountered. In the mid-2000s the US developed the Blue Force Tracker (BFT), which uses the same software as FBCB2, but operates over an L-Band satellite channel. Beyond line of sight communications provides a more comprehensive picture of the battle to all parts of the theatre.

Both systems operate a hub-spoke topology in which each vehicle connects to a Tactical Server Gateway (TSG) which maintains the Common Tactical Picture. The server regularly broadcasts updates to the picture to all connected nodes. For the latest BFT system the L-Band satellite channel has a 122 kbps forward link with a 2.8 kbps return link. This provides an update to the tactical picture within seconds.

Australia

In the late 1990s Australia's first experiments with digitisation launched the Battlefield Command Support System (BCSS) which was a ruggedised laptop hosting a suite of support applications. These applications included a digital battle map that displayed and tracked own force locations as well as geographic vector mapping detail of the terrain. Basic report messaging was also included. Over the last decade the BCSS has had a rolling program of improvements to the tools on offer and the technology used to deliver them.

More recently the BCSS network has been hosted on an EPLRS backbone to support better data throughput across the brigade formation. Despite its long history, the ADF recognised that the BCSS was better suited to planning functions than for use once the battle was joined. Therefore, in 2010 the ADF embarked on a new digitisation program to introduce a BMS that could be vehicle mounted and manage the distribution of tactical data.

The BMS is the central component of the Battle Group and Below Command, Control and Communications System (BGC3) that is being jointly delivered by the LAND 75 Phase 3.4, LAND 125 Phase 3A, and JP 2072 Phase 1 projects. It will incorporate a mobile, data capable communications system, and be able to exchange combat information through CNR links with BCSS and other land BMS. A Track Management System (TMS) will provide the capability to receive and exchange track data between various systems. TMS has adopted the Variable Message Format (VMF) message standard, and as such will be able to exchange information via VMF to BMS, AFATDS artillery direction systems, and any others that use the VMF format. TMS also includes strategic and intelligence feeds at the Secret level [1].

NATO

The principal European nations have all developed comprehensive digitisation programmes, which are in varying stages of maturity. In general they all provide much the same core functionality: BFT; a common tactical picture; the preparation and dissemination of orders and reports using both text and graphics; map and terrain analysis; and messaging and communications using email, chat and SMS. In some cases separate applications support specialist functions such as fire support or combat engineering, in others a separate system has been developed for this which is interoperable with the core system, particularly in terms of drawing on a common tactical picture. This is particularly true of fire support.

The UK digitisation experience has not been an entirely happy one. Some legacy projects provided automated C2 capability, but with elderly communications systems the UK adopted the “big bang” approach to modernise its whole C2 infrastructure. Having embraced the concept and acknowledged its centrality to future operational doctrine, the overall digitisation programme was closely integrated with the introduction of a new digital communications

(a) Operating FBCB2/BFT

(b) FBCB2

Figure 6.4.1: BMS Interfaces

system, Bowman. The prime contractor for this has been General Dynamics UK, which was awarded the contract in 2001after a long and painful gestation period which saw other contractors fall by the wayside [4].

India

The Indian Army's battle management effort is focussed around a BMS programme. The initiative is intended to be procured solely through local Indian companies, although a degree of partnering will be permitted to allow these firms to acquire relevant technologies internationally as and when required. The BMS aims to link infantry soldiers to higher command levels while providing a BFT capability and some measure of sensor fusion for both individual soldiers and other command elements. The BMS will carry voice, data and imagery traffic to individual computers used by soldiers, and computers used by other command elements in vehicles and at fixed-sites. Four BMS prototypes will need to be developed for mountain, desert, plain and jungle operations [5].

Malaysia

Malaysia is currently implementing a Network Centric Operations initiative through a partnership with Sapura, a Malaysian-based technology company. The first phase is a two-year study examining how to federate existing C2 systems used by the Malaysian Army. Future work may include the introduction of a new BMS along with the integration of Malaysia's SAKTI infantry modernisation architecture into the overall battlefield network [5].

6.4.3 Interoperability

Automated C2 systems have been around for some time both in the US and Europe, and it was well understood that a key requirement was to achieve interoperability between the systems. The US ABCS initiative was in response to the stove-pipe development of many of the support packages which prevented seamless integration of data. In Europe, the NATO forum produced what eventually became the Multilateral Interoperability Programme (MIP), the objective of which was to achieve command and control interoperability at reduced cost by developing and implementing technical standards agreed by NATO nations and prescribed by NATO. The aim of the programme was to identify the minimum set of specifications to be included within command and control systems, which would allow interoperability of systems between nations. The MIP specification "is a managed interface between command and control information systems. When incorporated into a system it enables interoperability of information between any other system that also incorporates the specification. Battlespace data is transferred as information. The meaning and context of the information is preserved across national boundaries precisely and without any ambiguity."

For the US there exists a similar issue when working with coalition forces. With the US firmly in the digital battle map world and relying on BFT to control forces and avoid fratricide, a coalition partner who is not on the network is also not in the tactical picture. The US system uses the Variable Message Format (VMF) to transfer data between applications. Therefore, other systems that also use VMF can usually achieve substantial interoperability with US forces, particularly the blue force tracking function.

6.4.4 Variable Message Format (VMF)

The Variable Message Format is a tactical data link scheme developed by the US to support land operations, including littoral operations and close air support. It is similar in concept to Link 16 in that it has a fixed message catalogue but each message can vary its structure within certain rules. The system has three main elements: the bearer, the message header and the message. These elements are defined in the following military standards:

- MIL-STD-188-220 Bearer Definition
- MIL-STD-2045-47001 Headers
- MIL-STD-6017 Message Set

Bearer Definition

MIL-STD-188-220 defines the physical, data link and network layers that may be used when connecting a Data Message Transfer Device (DMTD) over CNR. At the physical layer the standard defines the various modulation schemes that can be used, including FSK frequency spacings and PSK data rates. At the data link layer a set of FEC coding schemes are defined, including generator polynomials and convolutional encoding methods. At the network layer the standard defines a one-hop routing protocol to promulgate the network topology across the network to each node, without consuming excessive bandwidth. By mandating MIL-STD-188-220, VMF can operate over bearers ranging from HF/VHF CNR to UHF satellite communications. In contrast to Link 16, which uses a TDMA access method, the VMF media access protocol is a CSMA/CD method in which each node senses activity on the network and then broadcasts to all nodes when the medium is available. MIL-STD-188-220 defines several network addressing schemes that can be used to determine a stations eligibility for transmissions. VMF nodes can address messages at the data link layer, using IP addressing or using an operator address. There are also varying Types of Service which define a message profile for connectivity, speed of service, speed of recovery and reliability. Although MIL-STD-188-220 defines the interface between the DMTD and the transmission channel, in reality the DMTD is usually connected to a radio and the radio then provides the transmission interface. Therefore, a MIL-STD-188-220 compliant system depends upon the DMTD device and the selected radio.

Message Headers

The VMF message headers are defined in MIL-STD-2045-47001.

Message Set

The VMF message set is defined in MIL-STD-6017 and because it is incorporated into the US Army FBCB2 system, VMF is now the most widely used tactical data link. The catalogue of messages is expanding and covers the following functional areas:

- FAD 00 Network Control
- FAD 01 General Information Exchange
- FAD 02 Fire Support
- FAD 03 Air Operations
- FAD 04 Intelligence Operations/Imagery
- FAD 05 Land Combat Operations
- FAD 06 Maritime Operations
- FAD 07 Combat Service Support
- FAD 08 Special Operations

- FAD 09 Joint Task Force Operations Control
- FAD 10 Air Defence/Air Space Control

VMF has been implemented by the Australian military in its BMS programme and the Spanish air force in support of their F-18 fleet, with a UK implementation also under development. Many other European nations are also considering VMF for future development.

References

[1] Kym Bergmann. Battle management systems comes together. *Asia Pacific Defence Reporter*, October:12–13, 2012.

[2] R.D. Bowman. Improved performance for iintegrate voice and data in a tactical packet network. In *Military Communications*. IEEE, 1998.

[3] R.D. Bowman and H.B. Russell. Channel access for simultaneous voice and data transmission in a multiple-hop packet radio network. In *Military Communications*. IEEE, 1998.

[4] Giles Ebbutt. Digitisation of the battlefield. *European Security and Defence Press Association*, 2013.

[5] Thomas Withington. Middle managers: Battle management systems. *Asian Military Review*, 2014.

7 Vehicle Installation Issues

MODERN military fighting vehicles are being fitted with an increasing inventory of electronic equipment, including VHF/UHF/HF radios, BMS, GPS receivers, power re-charge facilities, crew interconnection systems and electronic countermeasures. Each of these devices brings increased capability to the vehicle crew, but each demand their share of the space, weight and power budget. Installing this portfolio of equipment brings its own special challenges to vehicle design engineers to ensure that each system can function within the close electromagnetic environment of a fighting vehicle. Communications security (COMSEC) brings an additional layer of complexity through the requirement to separate encrypted and unencrypted information. In this chapter we will explore some of these practical issues and consider how to manage them, or trade them against one another.

7.1 Radiation Hazard (RADHAZ)

The average mobile phone transmits a maximum of about 300 mW during a telephone conversation, yet there have been ongoing claims for years about the possible danger of excessive radio frequency (RF) exposure by using a handset against the side of the user's head[1]. By contrast, the antenna on a vehicle fitted with military radios might be one half to several metres away from the operator, but it can be transmitting 5, 10, 20 or even 50 Watts in the VHF/UHF range. Vehicle fitted HF radios can transmit up to 200 Watts. Exposure to RF emissions is a genuine health risk for radio operators and minimising this hazard should be an essential design consideration for any vehicle installation.

Some protection is offered by the fact that the operator is generally inside the vehicle, which can be approximated as a protective metal cage, while the transmitting antenna is outside. Nevertheless, on fighting vehicles, or on military trucks with a cupola, it is common for the crew to be exposed outside

[1] See http://www.arpansa.gov.au/mobilephones/mobiles1.cfm for a discussion of this issue.

the vehicle while transmissions are being made. For this reason the RF hazard must be assessed for each vehicle design that includes a communications suite.

7.1.1 National Standards

The relevant safe-working standard will vary from country to country. In Australia the standard is the *Australian Radiation Protection And Nuclear Safety Agency (ARPANSA) Radiation Protection Standard – Maximum Exposure Levels to Radiofrequency Fields – 3 kHz to 300 GHz*. The ARPANSA standard provides a range of Basic Restrictions that are considered to be adequate to ensure safe working conditions for personnel exposed to RF emissions. The ARPANSA standard lists the Basic Restrictions in terms of quantities such as Specific Absorption Rate (SAR), measured in Watts per Kilogram; Spatial Peak Specific Absorption (SA) in the Head within a 50 microsecond interval; and limb current density. However, the standard also recognises that these restrictions are difficult to measure in practice. Therefore, the Australian standard also lists a set of equivalent Reference Levels of more measurable quantities. The Reference Levels have been conservatively calculated such that if an environment satisfies the Reference Levels then it is considered to be compliant with the standard. For example, if the received power flux density at a location of interest *in the far-field* is below 10 Wm^{-2} then the level is considered to be safe for workers who are aware of radiation hazard risks.

The ARPANSA standard has been written around the assumption that measurements of the electric and magnetic field strength will be taken in order to determine compliance. The proximity of other metallic objects, particularly in the radio frequency near field, can have a significant impact on the distribution of electric and magnetic fields, making the effects difficult to calculate, so the only way to truly test the design is to perform field strength measurements after the system is built. Of course, prior to construction software modelling can also be performed to provide assurance that the design is likely to comply with the standard, but the model will require an accurate CAD model of the vehicle and specialist RF modelling software.

7.1.2 Example 1 - Far-Field Calculations

Usually, the communication antennae are mounted on the roof or cooling rack of a large military truck, or on the hull of an armoured fighting vehicle. Con-

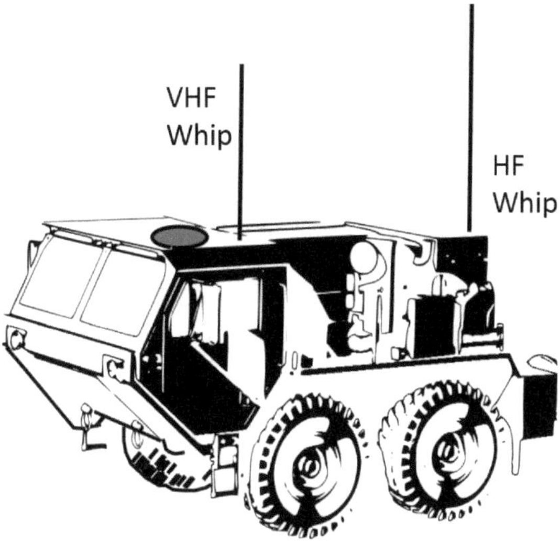

Figure 7.1.1: Military truck with HF/VHF antenna

sider a large military truck with both VHF and HF radios fitted in the cabin, the HF antenna mounted at the rear of the cooling rack and the VHF antenna mounted on the cabin roof, as shown in figure 7.1.1.

The geometry of the installation is illustrated in figure 7.1.2 and the problem parameters are listed in table 7.1, along with the calculation of the power flux density at the cupola. For calculations the weight of the soldier in the cupola is assumed to be 80 kg and the exposed area is 0.45 m^{-2}.

As a rough first estimate, it would appear that the geometry does not create a radiation hazard, since the power flux density received at the cupola is less than the maximum exposure defined in the standard. Unfortunately, it is not quite so simple. The Reference Levels can only be applied in the *far-field* of the antenna because they rely on the ratio between the electric and magnetic fields being constant at 377Ω. Therefore, we must test to see whether there are any operating frequencies at which the cupola falls into the *near-field.*

7.1.3 Near-Field vs Far-Field

Around a radiating antenna the behaviour of the electric and magnetic fields can be broadly separated into three regions; the near reactive field; the near radiative field; and the far field. Generally, the near reactive field boundary is considered to lay at a distance of approximately $\frac{\lambda}{2\pi}$ metres from the antenna,

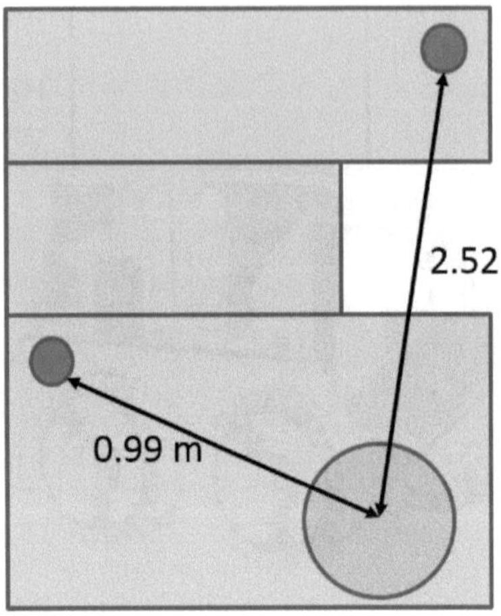

Figure 7.1.2: Roof geometry

System	Max Power (W)	Antenna Gain	eirp (W)	Distance to Cupola (m)	Received Power Flux Density (Wm^{-2})
VHF Radio	50	1.412	70.6	0.99	5.75
HF Radio	150	1.412	211.9	2.52	2.65
				Total	**8.40**
				ARPANSA Limit	10
				Margin	**1.6**

Table 7.1: Example HF/VHF installation

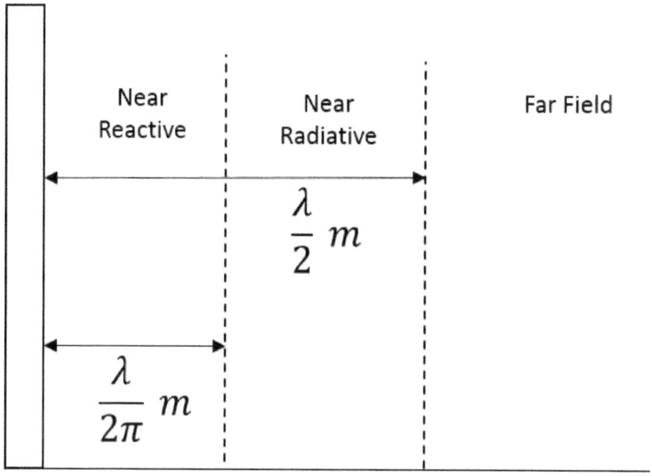

Figure 7.1.3: Near Field vs Far Field

where λ represents the wavelength of the emission. The far field is considered to start approximately $\frac{\lambda}{2}$ metres from the antenna, as shown in figure 7.1.3.

The behaviour of the electric and magnetic fields in the near-reactive field is complex and unpredictable. The Reference Levels provided in the ARPANSA standard have been constructed on the assumption that the ratio of electric field strength to magnetic field strength is constant, at approximately $\frac{E}{H} = 377\Omega$, which is true in the far field, but cannot be guaranteed in the near reactive field. The ARPANSA standard also acknowledges that field behaviour is somewhat unpredictable in the near radiative field, but claims that the ratio E/H is within 10% of 377Ω, so that the Reference Levels may be applied as long as the point of measurement is greater than $\frac{\lambda}{2\pi}$ metres. It urges caution when measuring close to the near reactive field boundary. The standard specifies that when the measurement point is within the reactive near field then power density calculations are not appropriate because the ratio of electric to magnetic field strength is not constant. Instead, the electric and magnetic field strength must be measured individually. In addition to the figures in table 7.1 there will be complex electric and magnetic field interactions in the near reactive field that are many times greater. Human capacitance represents another reactive component which will interact with these fields.

Given that the near field boundary is dependent upon wavelength, over the military HF/VHF bands (3-70 MHz) the radius of the near field boundary will vary from 15.9 metres down to 0.7 metres. Clearly the HF band and the low

VHF band present the greatest risk.

7.1.4 Example 2 - Near-Field Hazard

Using the the data in table 7.1, the frequency range in which the cupola is inside the near-field boundary of the VHF radio is 50 MHz and below. For the HF radio the cupola is in the near-field for all frequencies below 19 MHz. Since in the near-field the electric and magnetic reactive fields can be an order of magnitude larger than in the far-field, this rough calculation indicates that this arrangement poses a potential radiation hazard that should be properly modelled in software and then tested in the final design. A more detailed explanation of the near-field effects is included in Appendix 3.

7.2 Electromagnetic Interference and Compatability (EMI-EMC)

7.2.1 Overview

Electromagnetic interference is caused by one system to another system through the action of radiation, induction, conduction or capacitive coupling, as illustrated in figure 7.2.1. Each of these actions can cause spurious currents to form in the affected system, creating noise or disrupting the correct operation of the electrical system.

In general, a military radio is tested and certified against a military EMI standard, and similarly a vehicle is usually warranted by the manufacturer to function electrically without one component interfering with another. The difficulty comes through the need to integrate a sophisticated communications suite into the vehicle, which will draw vehicle power and radiate electromagnetic energy, and operate both systems successfully. Installing a communications suite into a vehicle requires careful attention to each of these risks because either system may be at risk. The vehicle may create a disturbance in the radio or BMS, or conversely, the communications equipment may induce electrical faults into the vehicle electronics.

The two principal methods of interference are conduction and induction. Conduction occurs when a signal from system A is transmitted along a wire that is subsequently connected to an input to system B. System B is then receiving interference from system A, which will usually manifest itself as degraded

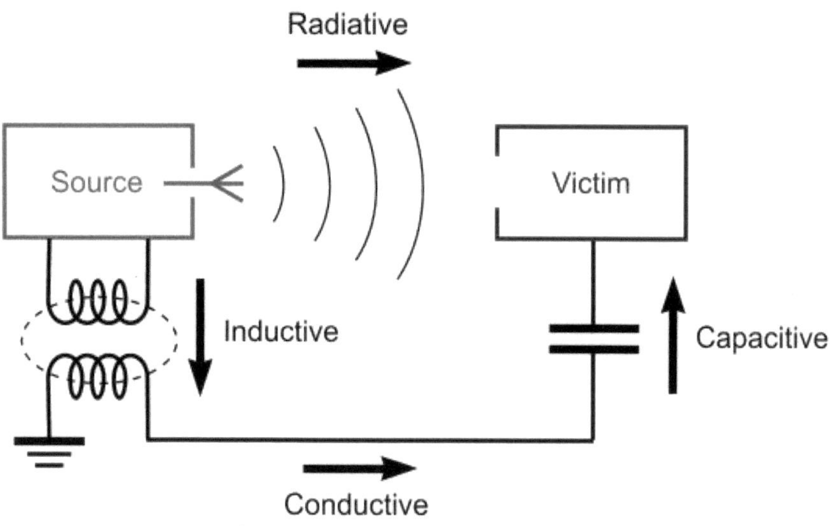

Figure 7.2.1: EMI Modes

performance. In military vehicles the conduction risk is typically through the common vehicle power supply and occurs when signals from an electronic component are allowed to travel back along the power supply cable and then forwards into a second electronic system, as shown in figure 7.2.2.

Induction is the effect of radiation causing induced currents to flow in the second system and requires no direct connection from one system to another. There are many examples of induction in everyday life, such as hearing a second telephone conversation whilst speaking to somebody else. This effect is called cross-talk and occurs when the electrical signals from one telephone wire radiate from the wire and are induced into a nearby second telephone wire. This is the reason that many security standards demand physical separation of unencrypted (red) voice or data wires from transmission lines (black).

Within military vehicles fitted with communications equipment induction can occur through several mechanisms. Clearly the radios broadcast significant energy in the HF/VHF/UHF frequency bands so it is essential that the vehicle is not susceptible to electromagnetic interference at those frequencies. Conversely, the radios are receiving in the same bands, so it is essential that the vehicle does not radiate significant energy in those bands. Radio equipment also operates at one or more intermediate frequencies which may be outside of the operating band, so the vehicle must be tested to ensure that it does not

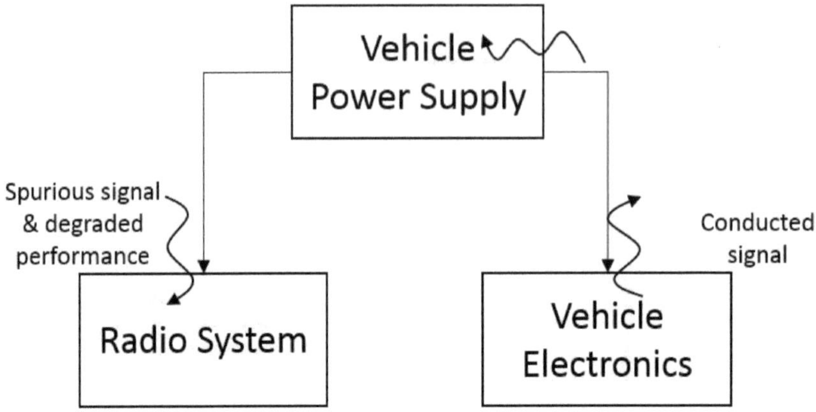

Figure 7.2.2: EMI conduction

generate energy at the intermediate frequencies. This particular risk is usually mitigated because most military radios comply with the electromagnetic susceptibility requirements of MIL-STD-461, so they are protected against interference through shielded enclosures and other design practices.

7.2.2 Standards: MIL-STD 461 (DEF-STAN 59-411)

The two principal standards for avoiding electromagnetic interference are MIL-STD-461F and the UK DEF-STAN 59-411. These standards specify performance levels for military equipment against specific test methods to test for conducted emissions, radiated susceptibility, and radiated emissions. We will consider MIL-STD 461 in further detail.

Conducted Emissions CE 101, CE 102, CE 106

MIL-STD-461 sets out two test regimes for conducted emissions along power leads: *CE101, conducted emissions, power leads, 30 Hz - 10 kHz,* and *CE102, conducted emissions, power leads, 10 kHz - 10 MHz.* The 28 Volt power system in a military vehicle is usually defined by MIL-STD-1275 which specifies power consistency and limits voltage ripple. Usually a communications item will be certified by a third party testing organisation at the time of design and manufacture to be compliant with these conducted emissions requirements, so it would be unusual for the radio systems to introduce a conducted emissions problem into the power supply of the vehicle.

CE 106 is a test of the emissions from an antenna port which checks for spurious emissions or excessive harmonic energy by measuring the emissions along an antenna cable.

Radiated Susceptibility RS 101, RS 103, RS 105

Radiated susceptibility testing ensures that the equipment under test is robust enough to continue to operate in the presence of moderate radiated signals. RS 101 examines equipment performance in the spectrum from 30 Hz to 100 kHz, RS 103 examines the spectrum from 2 MHz to 40 GHz and RS 105 examines the equipment response to high transient peak bursts of electromagnetic energy. The aim of the test is to ensure that *"The EUT shall not exhibit any malfunction, degradation of performance, or deviation from specified indications..."* when it is irradiated with a signal at the test frequency and at the specified level of intensity.

For military equipment the standard is generally higher than for equivalent commercial equipment because deployed military nodes generally have a high concentration of electromagnetic emitters. Susceptibility testing routinely discovers conditions under which commercial equipment, that has been adapted for military use, responds abnormally when irradiated at a particular frequency because the commercial imperative did not demand such protection during the design phase.

Radiated Emissions RE 101, RE 102, RE 103

Radiated emissions testing is designed to ensure that a new piece of equipment does not emit excessive energy at unexpected frequencies. Clearly a radio is expected to radiate when transmitting, but it should not emit at frequencies outside the transmit range. Similarly, when in receive mode there should not be spurious energy emissions. From the vehicle perspective, it should not radiate energy beyond the standard limits simply through turning on the air conditioning or the windscreen wipers. Measuring a vehicle's radiated emissions requires a large anechoic chamber (or often a semi-anechoic chamber) in which the vehicle is surrounded by receive antenna horns. The vehicle is turned on and the various electronic devices are engaged while measurements are taken of the electromagnetic spectrum.

MIL-STD-461 defines three tests for radiated emissions. RE 101 examines the magnetic field radiated from 30 Hz to 100 kHz; RE 102 examines the electric

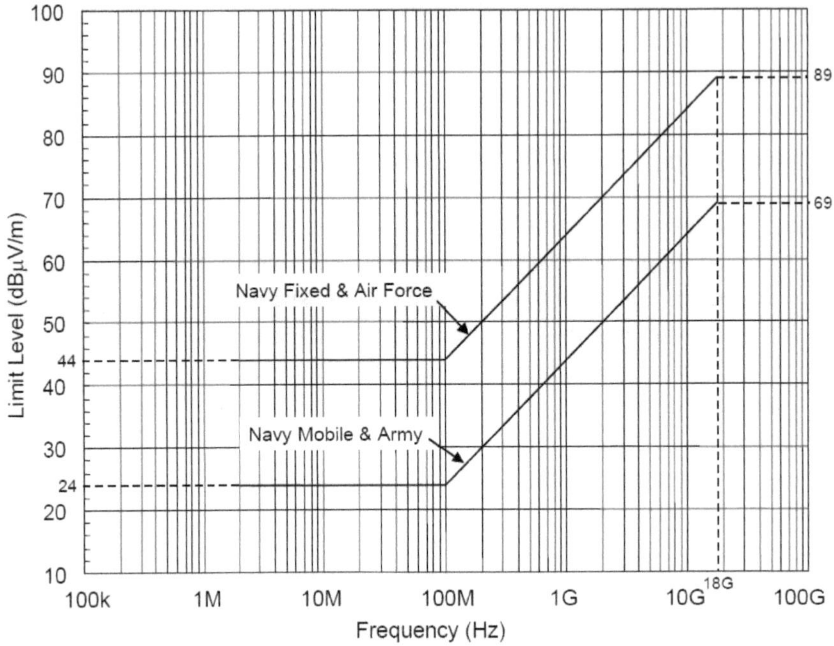

Figure 7.2.3: MIL-STD-461F RE 102 emissions limit

field radiated from 10 kHz to 18 GHz; and RE 103 examines specifically antenna outputs to check for spurious or harmonic emissions from 10 kHz to 40 GHz. RE 103 is an alternate to CE 106, although CE 106 is the preferred approach. The permissible emission limit is shown in figure 7.2.3.

7.2.3 Example Vehicle

Most vehicles exhibit their strongest electromagnetic emissions in the low frequencies below 1 MHz, mainly due to mechanical rpm effects in the alternator and other components. Figure 7.2.4 shows an emissions performance graph for an example military vehicle across the frequency range 1 kHz to 1 GHz. Significant activity can be seen under 100 kHz, but the emissions reduce their intensity significantly after that. In the communications band there is some emissions activity in the vehicle, but it is well below the requirements of RS 102 from MIL-STD 461, which is also marked on the graph. In the higher frequencies the vehicle emissions generally drop away and are less likely to

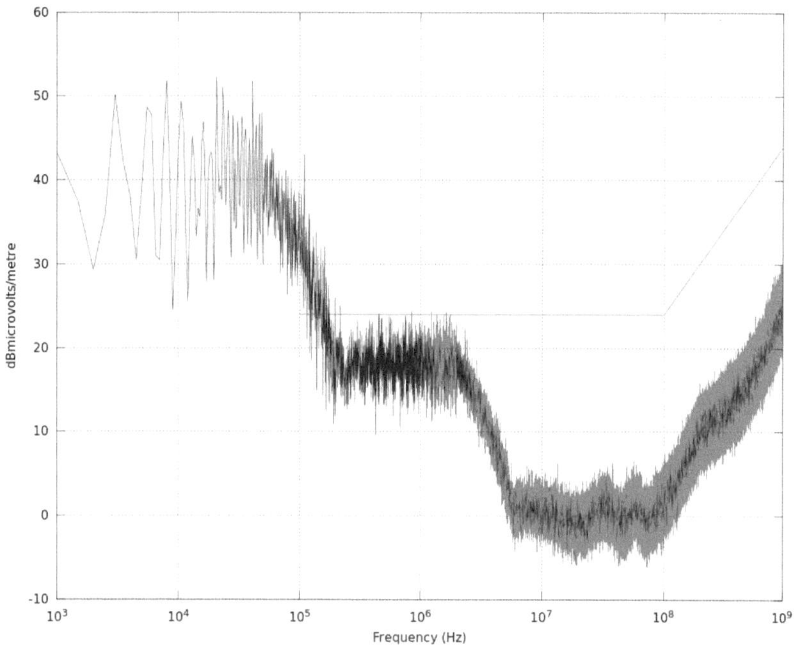

Figure 7.2.4: Typical vehicle emissions performance

cause an electromagnetic interference problem.

7.3 Physical Restrictions

7.3.1 Space claims

One of the very practical aspects of integrating a communications suite into a vehicle is ensuring that there is enough space in the cabin to mount all of the equipment. In modern military vehicles the claims for space are always increasing, while the cabin size doesn't seem to change. Communication systems must compete for space with weapon systems, navigation systems and crew comfort systems.

A simple installation might be two combat net radios and a BMS system. However, each radio will also need a shock mounting tray and a vehicle power

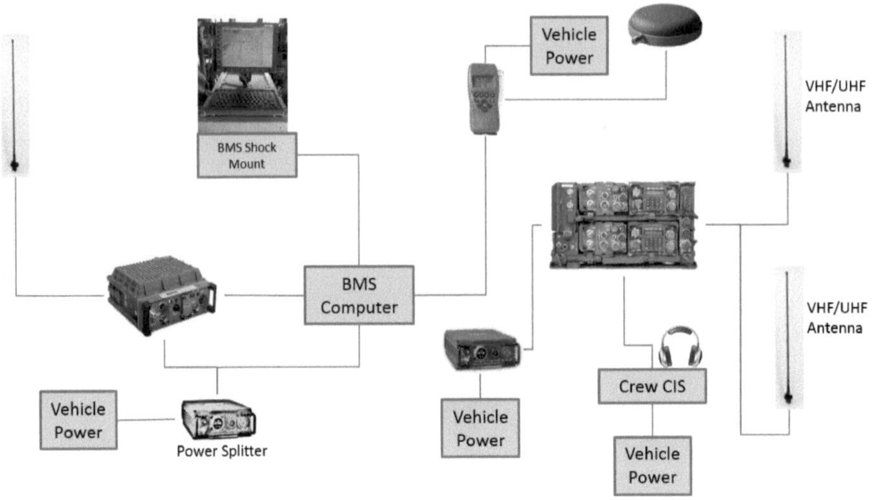

Figure 7.3.1: Typical C4I vehicle architecture

supply that also needs to be mounted in the cabin. The antenna cables will need to be fed through to the roof, but if it is an armoured vehicle then an approved armour pass-through is required for the cables, which may need a design approval. The BMS will be a ruggedised computer and a display. Sometimes they are integrated in the same hardware, or in other cases the computer is a separate item and a thinner display is used by the crew. In either case, the BMS screen needs to be mounted near the front of the vehicle and within reach of the driver. If the computer is separate then it is another box that needs to be fitted somewhere in the cabin. Many BMS also use a GPS input, so the GPS antenna will also need to be mounted and a cable input accommodated. If a Defense Advanced GPS Receiver (DAGR) is used then it will also need to be mounted somewhere.

A typical vehicle communications architecture is shown in figure 7.3.1. Adding each additional item generally produces a non-linear increase in the complexity of the installation.

7.3.2 Security

Most armies demand secure encrypted communications for their forces, so communication security devices will also need to be fitted into the cabin. Some radios offer in-built encryption, but this is not available in every product.

Security can also influence the location of equipment within the cabin, since normal security protocols will separate "red" and "black" communications to avoid accidental cross-talk or unexpected emissions. These considerations may limit the space options available for installing the communications suite from both an equipment perspective and also a cabling perspective.

7.4 Co-site interference

EMI testing can identify unwanted emissions that may interfere with the correct operation of the communications equipment, but there is another interference issue that can occur when multiple radios are installed in a vehicle. Co-site interference is the emissions from system A being received by system B simply because of their close proximity to one another. Under normal operations, when each antenna lays in the far-field of the other system, the normal operation of the radios should preclude interference from one another through sensible frequency allocation. By keeping the frequencies reasonably well apart, any harmonics generated by system A, that might interfere with the portion of spectrum being used by system B, will generally be of such a low power level that the adjacent channel rejection designed into the system B radio will be able to keep the spurious emission out of system B.

Unfortunately, when antennae are placed in close proximity to one another, further complications can arise. We have already encountered difficulties with the near-field because the electric and magnetic fields in the near-field can be an order of magnitude greater than in the far-field. On a vehicle the antenna locations are limited and it may be that at certain parts of the operating spectrum the near-field of the system A antenna could overlap the antenna of system B. In this case the reactive coupling between the transmission and the system B antenna can induce spurious noise currents into the radio, producing interference and degrading the communications channel. For each additional radio added to the vehicle the problem increases in complexity since it moves from a "one on one" problem to an "N on one" problem. As shown in figure 7.4.1, additional effects can occur when system A interferes with system B, producing an unexpected intermodulation product, which then affects system C. Yet, individually, or "one on one", there may be no interference. Solving a co-site interference problem on a vehicle installation is difficult because there are few options available. Generally the space constraints limit antenna placement so other electronic solutions may be required, such as adding addi-

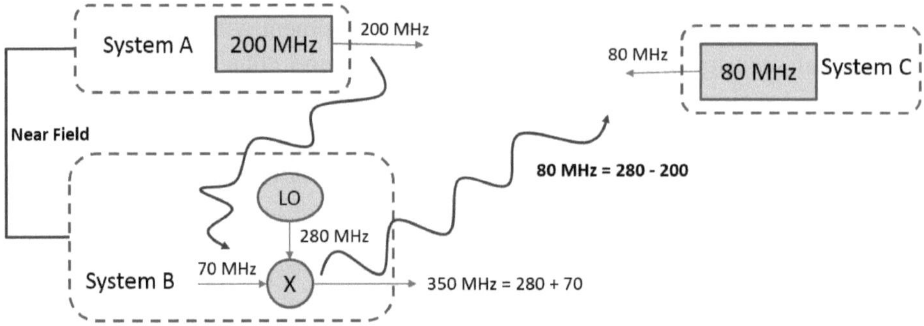

Figure 7.4.1: N-on-1 co-site interference

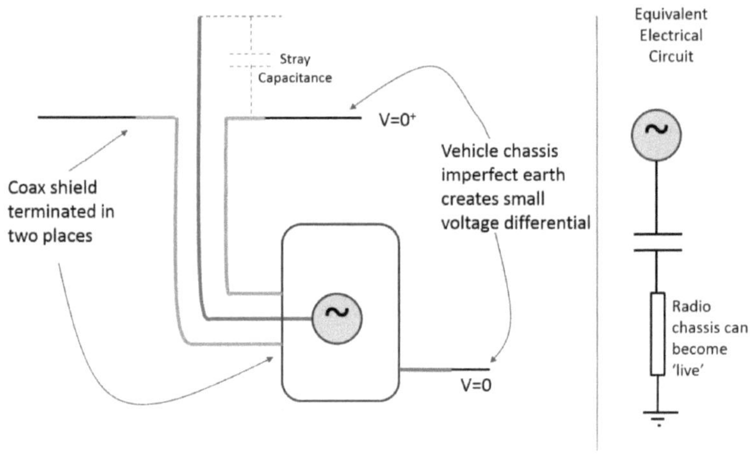

Figure 7.4.2: Common mode noise due to imperfect earth plane

tional filters into the cabling or antennae systems in order to attenuate specific troublesome frequencies.

Another electromagnetic challenge with vehicle installations is the presence of common mode noise along the outside of the coaxial shield of the antennae cables, as shown in figure 7.4.2. This noise is generated due to the imperfect ground plane presented by the vehicle chassis, which induces stray capacitance between the antennae and the vehicle. Common mode noise can create significant interference effects within the communications equipment and will degrade the transmission performance of the equipment. Typically, an inductive filter, called an RF choke, is placed in the cable, near the antenna termination, to attenuate the common mode signal.

7.5 Vehicle Power Budgets

7.5.1 Overview

The installation of a communications suite into a military vehicle has an inevitable impact on the power budget of the vehicle. The extent of the impact is governed by the number of elements in the communications suite and the type of radios fitted. A VHF/UHF CNR typically has a maximum power output of 50 Watts whereas a HF CNR may have a vehicle amplifier that can transmit 150 Watts. In assessing the power performance of the vehicle it is important to understand the base load and then to estimate the peak and average load increase introduced by the communications suite. Of course, the communications suite is just one claim on vehicle power. Increasingly there are other claims from such services as detection systems, silent watch, sensors, reconnaissance tools and weapons.

7.5.2 Power Input

The power input is provided by the vehicle alternator which the manufacturer will specify across a temperature range and for a motor rpm range. Figure 7.5.1 is a typical curve for a high output (200 Amps) alternator. Note that the 200 Amps is only available when the engine is running hard; at idle, which is around 800 rpm in this example, the alternator only delivers 120 Amps. In calculating the power budget it is important to keep the mission profile in mind, since if the radios will be used extensively while the vehicle is stationary then the lower output figure should be considered.

7.5.3 Base Load and Communication Load

The base load of the vehicle should be provided by the vehicle manufacturer and should account for every electrical fitting on the vehicle. It is then helpful to consider several operational scenarios for the vehicle, such as a battle mission and an admin move, or a mission in hot weather and a cold weather mission. In calculating the power profile the operational profile is important because not every electrical item will be turned on simultaneously. For example, the engine heating (for freezing weather) and the air conditioning are unlikely to operate at the same time, yet each has a significant power requirement. Without considering the operational profile the power budget can become unrealistically high.

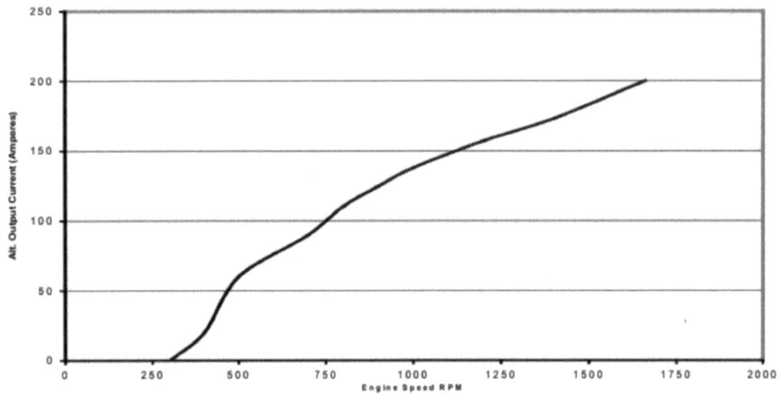

Figure 7.5.1: Example alternator curve

Similarly, the operational profile of the communications equipment should also be assessed in determining the power requirements. For example, consider a VHF CNR with a nominal power output of 50W. In reality, the transmitter will not be completely efficient so let us assume a conservative 66% efficiency. Therefore, the power drawn by the radio may be as high as 75W in order to produce the 50W output. If the vehicle has a 24 V power system then that equates to a current requirement of $75 \div 24 = 3.125$Amps. However, in operations the radio may only transmit for a total of ten minutes each hour, producing an average power requirement of $3.125 \times \frac{10}{60} = 0.52$ Amps.

Of course, when the radio is not transmitting it is in standby mode and consumes, perhaps, 20% of the nominal power, or 10W. Therefore, the standby current consumption is given by $(10 \div 24) \times \frac{50}{60} = 0.35$ Amps. Therefore, the total current drain due to the VHF CNR is approximately $0.52 + 0.35 = 0.87$ Amps.

The radio manufacturer can usually supply accurate figures for power consumption to assist with the modelling. All of the communication system elements should be modelled in the vehicle power budget to ensure that there is sufficient power margin for the vehicle to operate in each of its mission profiles. This work will also support calculations for the maximum duration of any silent watch facility required by the vehicle. Ideally, a margin of 30% would be preferable to support the installation of further electronic devices.

7.6 Vehicle Electronics (Vetronics)

7.6.1 Background

"Vetronics" is a contraction of vehicle electronics and represents a body of work over the last decade to define an open standards architecture for the electrical system in military vehicles. The work originated in the UK through the Vehicle Systems Integration (VSI) Advanced Research Program, but through the Military Vetronics Association interest has extended across many European nations and to some extent the United States[2].

As military vehicles become increasingly complex and expensive, the drive for cost effectiveness requires that these vehicles should have longer in-service lifetimes than current vehicles. Changing operational commitments and scenarios coupled with new and varied threats suggest that platform reconfiguration, or technology insertions, will be required to meet the changing demands. These changes must be achieved at minimal cost and in short time scales, resulting in the need for platforms equipped with an infrastructure based on open architectures.

Current fielded platforms have generally been designed with little thought to changes during the lifetime of the platform. Providing an infrastructure to accommodate upgrade capability has generally been deleted from the production build standard to save on initial procurement cost, with the result that retrospective integration on current vehicles is very expensive and time consuming. Open architectures, based on VSI standards & technologies, will promote cross-fleet commonality which in turn will lead to savings generated by an anticipated reduction of costs.

Building a platform based architecture can be done in several ways, which are each discussed in the Vetronics Standards and Guidelines Issue 3 [1].

Single Common Architecture

The first option is to employ a single common architecture, creating a single data network, that can accommodate all requirements from low bandwidth audio data through to high bandwidth digital data. In this case all subsystems are either connected via a dedicated node to the network or a number of subsystems are grouped together and connected via shared node. This approach

[2] http://www.milva.org/index.php/about-milva

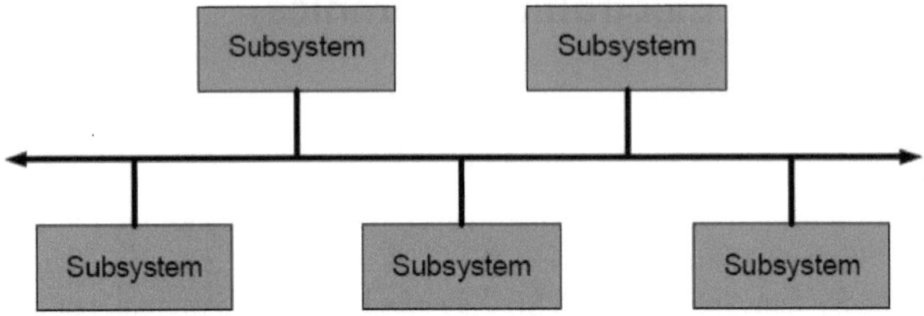

Figure 7.6.1: Single common Vetronics architecture

creates a single standard method of connecting into the vehicle architecture, which makes it a simple solution.

The advantage of this approach is that only one type of network interface need be defined for all subsystems. This provides commonality between subsystems and across platforms, which eases the integration burden. However, the approach has several disadvantages. The network technology must be chosen to meet the both the highest and lowest data rates required and is therefore likely to be a compromise. A high data rate interface will be unnecessarily expensive for low rate applications and the cost of the interface may be considerably greater than that of the function to be integrated. Also, the network technology must be carefully chosen to ensure both industry support and longevity. Finally, data transfer via a single network would compromise safety critical functions, since changes to the system of a non-safety critical nature may impact those that are safety critical.

Segregate by Data Type

An architecture partitioned by data type can be constructed by using technologies appropriate to bandwidth requirements. In this case, high bandwidth subsystem data exchange; low bandwidth subsystem data exchange; and high bandwidth digital image data would all be implemented separately using technologies appropriate to their requirements.

The advantage of this approach is that the network technology can be selected to match data requirements, thus avoiding the cost of over specification. Cross fleet commonality can be achieved by specifying a particular technology for a data type, regardless of platform type or variant. A disadvantage is that multiple network technologies, matched to data types, will require bridges or

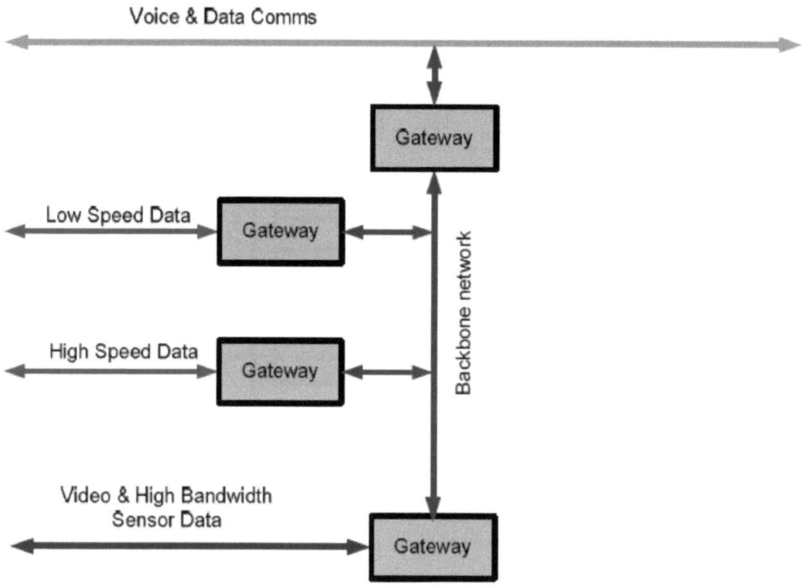

Figure 7.6.2: Segregation by data type

gateways wherever there is a need to transfer data between network elements. The safety critical functions would still be compromised, since there is no isolation between safety critical and non-safety critical changes to elements of the system.

Segregate by Function

The preferred architecture is to segregate the platform by functions, grouping similar functions together as illustrated in figure 7.6.3. This approach allows low data rate functions to be interfaced to the vehicle using a standard interface across all vehicle types. Higher data rate functions, such as video streaming can also be interfaced to the vehicle in a standard way across all fleets, albeit a different technology to the low data rate equipment. Approving several different interfaces is a compromise between expense and simplicity. A single standard interface for all equipment is simple, but could result in an expensive solution for some low data rate systems.

The standards and guidelines recommend protocols for the various platform networks and also provide a set of metrics which can be used to assess whether

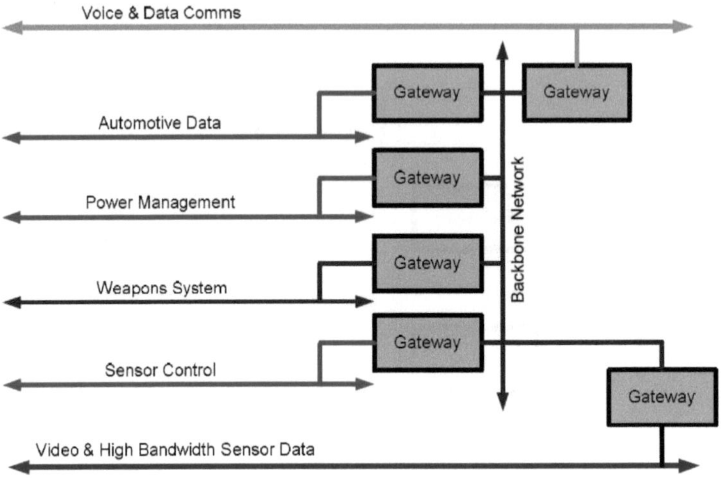

Figure 7.6.3: Vetronics preferred architecture

an architecture is vetronic compliant.

A major output of the VSI programme has been development and publication of the Vetronics Infrastructure for Video Over Ethernet (VIVOE) Defence Standard as Def Stan 00-82 Issue 2 [3]. The standard describes the mechanisms and protocols that shall be employed to implement the distribution of digital video over an Ethernet infrastructure within Vetronics systems.

7.6.2 Vetronics and BMS

Establishing an electronic architecture for military vehicles will create much more potential for "plug and play" system procurement. At the moment, system integration is a major cost driver in vehicle projects as disparate vendors provide equipment with a range of interface requirements. For example, France's "Scorpion" development programme, is a comprehensive project addressing both equipment and doctrine, a key element of which is digitisation. Its first, intermediate objective, is to enhance the sharing of the tactical situation and all aspects of battle management through the replacement of SIR-SIT-MAESTRO systems by SI/CS v0 (Système d'Information et de Combat SCORPION). This will be based on a single technical solution with an open architecture and a set of core services, with the specific functionalities provided as applications available as required.

The second phase will be the fielding of SI/CS v1. This will allow collaborative functions such as the management of direct and indirect fires and will benefit from the introduction of vehicle electronic architectures. This vehicle digitisation enables Vetronics to be integrated with the BMS, providing automatic updates of vehicle states and the aggregation of logistic data such as fuel states and is true of all modern vehicle programmes. It also allows the easy integration of vehicle mounted sensors such as laser range finders and electro-optics, as well as fire control systems [2].

References

[1] R.M. Connor. Vetronics standards and guidelines - issue 3, July 2009.

[2] Giles Ebbutt. Digitisation of the battlefield. *European Security and Defence Press Association*, 2013.

[3] Def stan 00-82: Vetronics infrastructure for video over ethernet (viveo) issue 2.

8 Electronic Warfare

ONE of the unique characteristics of battlefield communication systems is that they must be designed to operate in a hostile electromagnetic environment. Whilst all systems must cope with unwanted interference, military communications will be subject to deliberate electronic attack, or jamming, that is designed to disrupt communications. The victims of jamming are always receivers, and deliberate jamming is a competition at the receiver between a legitimate transmitter and the jamming transmission. The jammer wins if it is able to disrupt communications to the point that information can no longer be shared between the legitimate parties. The jammer fails if the operators of the transmitter and receiver are able to cope with the jamming transmission and still share information between one another.

Jamming signals are subject to the same sorts of propagation losses that the legitimate signal suffers, so modelling the jammer is the same as modelling the legitimate transmitter. However, for normal communication links we are concerned with the signal to noise ratio, $\left(\frac{S}{N}\right)$, whereas in a hostile situation the jamming signal acts like an additional noise signal, and therefore we are interested in the ratio $\frac{S}{J+N}$. Usually the jammer power is much greater than the thermal noise power so it is common to consider only the ratio $\frac{S}{J}$, or more usually in the literature the jamming ratio is expressed as $\frac{J}{S}$.

8.1 Jamming Ratio $\left(\frac{J}{S}\right)$

8.1.1 Basic Scenario

Figure 8.1.1 shows the geometry of a typical electronic attack scenario. In this example we assume that the transmitter and receiver are typical vehicle mounted narrowband Combat Net Radio (CNR) with a channel bandwidth of 25 kHz and power output of 20 Watts. The VHF jammer is further away from the receiver than the signal transmitter, but is generally a high power device with typical power output of 500 Watts or greater. In the simplest

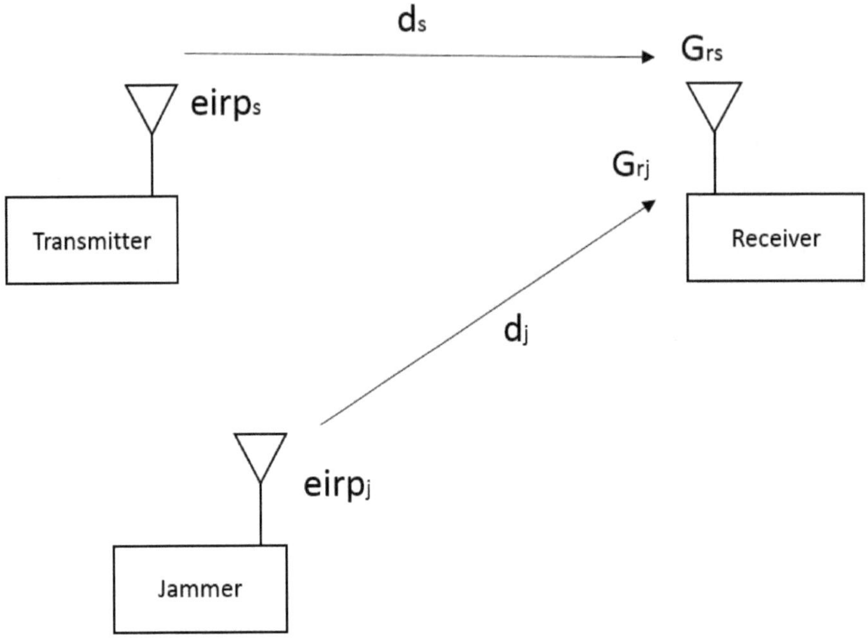

Figure 8.1.1: Typical electronic attack scenario

scenario the bandwidth of the jamming signal occupies the same bandwidth as the communications channel, which in this case is 25 kHz.

If we assume that free space loss is the principal loss factor then the power of the jamming signal at the receiver, J_r, will be

$$J_r = \frac{eirp_j G_{rj}}{4\pi d_j^2}$$

and the power of the transmitted signal, S_r, at the receiver will be

$$S_r = \frac{eirp_s G_{rs}}{4\pi d_s^2}.$$

Therefore, the jamming ratio will be

$$\frac{J}{S} = \frac{eirp_j d_s^2 G_{rj}}{eirp_s d_j^2 G_{rs}} \tag{8.1.1}$$

Parameter	Value
EIRP$_s$	14.5 dB
d$_s$	10 km
G$_{rs}$	1.5 dB
EIRP$_j$	37 dB
d$_j$	30 km
G$_{rj}$	1.5 dB

Table 8.1: Scenario parameters

or expressed in decibels as

$$\frac{J}{S} = EIRP_j - EIRP_s - 20\log\left(d_j\right) + 20\log\left(d_s\right) + G_{rj} - G_{rs}. \quad (8.1.2)$$

Note that the gain of the receiver in the direction of the jammer can be significantly different to the gain of the receiver in the direction of the transmitter, depending upon the type of antenna at the receiver. If an omni-directional dipole or monopole is used then G_{rs} and G_{rj} could be the same, but if a directional antenna is used, and oriented towards the legitimate transmitter, then the jammer is forced to attack through an antenna side lobe. The gain in the side lobe direction may be 20 dB below the gain along the bore sight, making the jammer's job much more difficult.

For analogue signals a jammer needs to achieve a J/S of between 0 to 40 dB (typically 10 dB) in order to disrupt communications. This is because the human operator has a significant capacity to work through a jamming signal, effectively "filling in the gaps". For digital signals a J/S of 0 dB for about one third of the transmission time will generally disrupt communications by corrupting sufficient bits that the the system is unable to recover. Hence, digital systems are more susceptible to jamming than their analogue counterparts [1].

8.1.2 Example Scenario - Analogue Communications

Table 8.1 lists the parameters for the tactical scenario described above. For this example the transmitter and receiver are assumed to be using a whip antenna with a gain of 1.5 dB, whereas the jammer is using a directional antenna, such as a Yagi-Uda array, with a gain of 10 dB.

Hence,

$$\begin{aligned}\frac{J}{S} &= EIRP_j - EIRP_s - 20\log(d_j) + 20\log(d_s) + G_{rj} - G_{rs} \\ &= 37 - 14.5 - 89.5 + 80 + 1.5 - 1.5 \\ &= 13 \text{ dB}\end{aligned}$$

which is likely to successfully disrupt a standard narrowband analogue voice channel.

8.2 Basic Electronic Counter Measures

Military signallers are taught to work through jamming wherever possible. The idea is that if the jammer cannot sense any change in operations then he may conclude that the jamming is ineffective and stop, in order to try a different approach. This is coupled to the reality that a high powered jammer presents an electromagnetic beacon on the battlefield, so it cannot transmit indefinitely. Jammers need to transmit and then move to a new location, or preferably transmit on the move. However, transmitting on the move whilst deploying a large, high-gain antenna is often not practical.

However, if the signallers are unable to work through the jamming then equation 8.1.2 suggests four basic approaches to countering the effects of deliberate jamming: shorten the distance between the transmitter and receiver, force the jammer to move further away, increase the signal transmitter power, or use a directional antenna at the receiver.

8.2.1 Path Distance

All other things being equal, equation 8.1.1 tells us that $\frac{J}{S} \propto \frac{d_s^2}{d_j^2}$. Therefore, to reduce the jammer to signal ratio one can either reduce the signal distance between the transmitter and receiver, or force the jammer to move further away. For an analogue scenario, if we assume that operators can work through the jamming if $\frac{J}{S} < 10$ dB then figure 8.2.1 shows the change in distance that would sufficiently reduce the ratio. The first figure shows that if we could force the jammer, through aggressive patrolling or targeted munitions, to move a further twelve kilometres away then the effect of the jamming would become manageable. Alternately, if we could move the receiver or transmitter three kilometres closer then the strength of the received signal would be sufficient for the operators to continue working.

8.2.2 Increased Signal Power

A second countermeasure to consider is increasing the power of the transmitted signal. Modern tactical radios have multiple power options, so it may be possible for the transmitter to increase the signal power sufficiently to overcome the jamming effect. The jamming ratio is directly proportional to signal power, so every dB increase in eirp_s directly reduces the $\frac{J}{S}$ ratio. For this example a 3 dB gain is required, or a doubling of the transmitter power from 20 Watts to 40 Watts. A typical tactical radio vehicle installation can offer power settings of 20 Watts and 50 Watts, so moving to the higher power setting may solve the problem in this scenario.

8.2.3 Directional Antennae

In this scenario the transmitter and receiver are combat net radios which usually use monopole whip antennae which have an omni-directional radiation pattern. As a result, the antenna gain in the direction of the jammer is the same as the antenna gain in the direction of the desired transmitter. However, if a directional antenna can be used instead of a monopole then it will be possible to reduce $\frac{J}{S}$ in all cases except when the jammer is aligned with the bore of the antenna.

Figure 8.2.2 shows the antenna radiation pattern for a ten element Yagi antenna with the off-axis gain displayed relative to the bore sight gain. Using this antenna, if the jammer were to be 30° of the bore then G_{rj} would be 10 dB lower then G_{rs}, which would be sufficient to defeat the jamming effect. At 40° there is a null in the pattern that would severely attenuate the jamming signal by around 30 dB and at 50° the jammer would be entering through a side lobe which would result in attenuation of 20 dB less than in the direction of the desired signal.

8.3 Electronic Attack of Digital Communications

Digital communication systems are more susceptible to jamming than their analogue counterparts because there is no human available to "adapt" to the

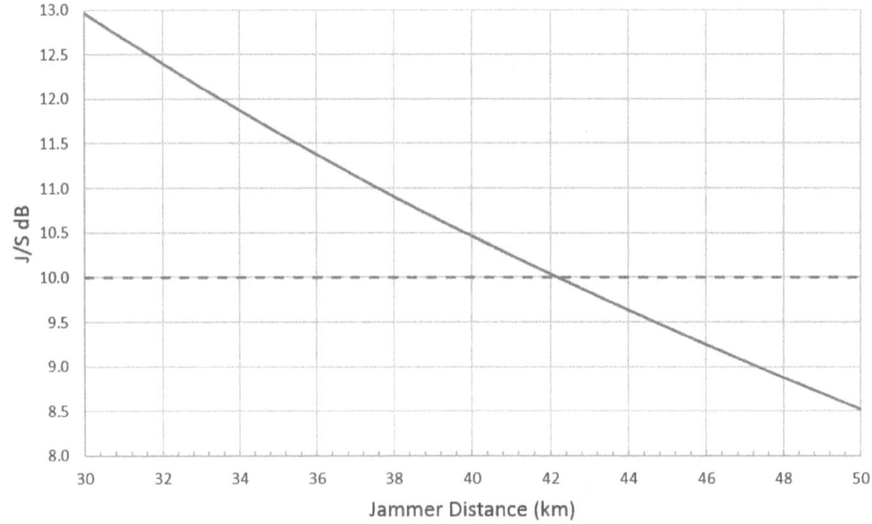

(a) Changing jammer distance

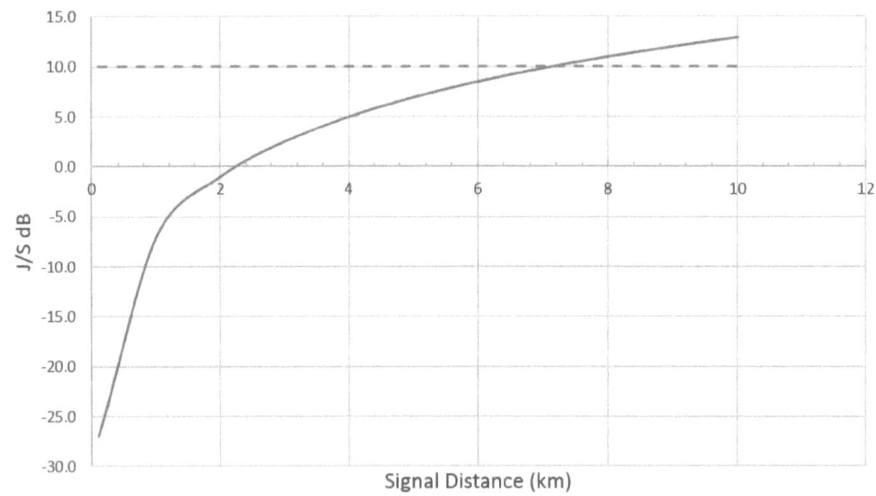

(b) Changing signal distance

Figure 8.2.1: Effect on $\frac{J}{S}$ by changing distance

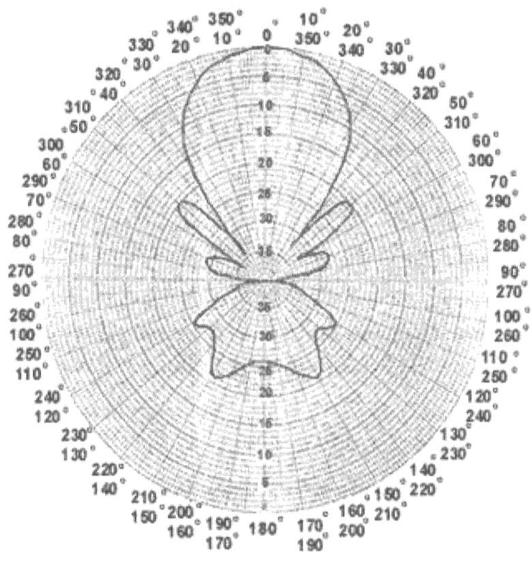

Figure 8.2.2: Antenna radiation pattern - ten element Yagi

jamming effect. Figure 8.3.1 illustrates that for coherently detected BPSK modulation, even a modest 10^{-5} probability of bit error requires an $\frac{E_b}{N_0}$ of 9.4 dB. In a hostile electromagnetic environment the jamming signal becomes the dominant noise source and a $\frac{J}{S}$ ratio of 0 dB is equivalent to an $\frac{E_b}{N_0}$ of 0 dB in a narrowband channel with spectral efficiency of 1 bps/Hz. If $\frac{J}{S}$ =0 dB in this example then the bit error rate will be around one in ten bits. Generally there is no advantage in pushing the jammer ratio any higher than 0 dB, because the increase in bit error rate is asymptotically approaching 50%. That is, for an extra 1 dB increase in the ratio there is very little increase in the error rate.

To make things even worse, if the signal is unreadable for, in the short term, around one third of the time then it is considered useless. This means that the jammer can disrupt a digital signal if the $\frac{J}{S}$ is 0 dB for a duty cycle of around 33%, compared with a positive $\frac{J}{S}$ for 100% of the time for analogue communications [1]. This 33% disruption must occur within the syllabic rate for speech or its equivalent in digital data. This weakness can be exploited by a jammer against even a broadband signal. Instead of attacking the entire bandwidth, which dilutes the power density, the jammer can use partial band jamming to attack only one third of the bandwidth. This improves the jammer's power density and still has a reasonable probability of disrupting digital communications.

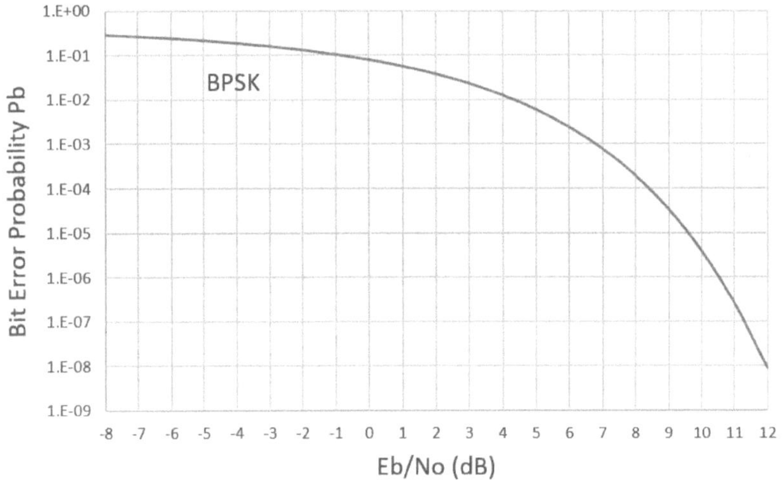

Figure 8.3.1: $\frac{E_b}{N_0}$ vs P_b for coherently detected BPSK

However, it is important to note that the point at which the $\frac{J}{S}$ ratio of 0 dB is measured is at the point in the receiver where the digital waveform is recovered from the RF modulation. Spread spectrum schemes introduce additional processing gain so that the $\frac{J}{S}$ ratio can be reduced prior to demodulation, and FEC coding can further recover bit errors after demodulation.

8.4 Advanced Electronic Counter Measures

Information is recovered after demodulation, so the critical J/S ratio is at the point of demodulation. For narrowband PSK or FSK schemes there is no advantage gained through demodulation, so the J/S prior to the receiver is the same as the J/S after demodulation. However, when spread spectrum techniques are included in the modulation scheme they introduce a processing gain that can directly improve the system error performance in the presence of deliberate electronic attack. Shannon's capacity formula provides some insight into why this can be achieved. In a system where the jammer power is the dominant noise source Shannon's formula can be expressed as

$$H = B\log_2\left(1 + \frac{S}{J}\right)$$

where H is the channel capacity and B is the channel bandwidth. We have seen previously in equation 5.3.2 that when $\frac{S}{J} \ll 1$, or the jammer is strong compared to the signal, then $H = \lim_{B\to\infty} \frac{S}{J_0}\log_2 e$. In other words, Shannon's Law tells us that even in the presence of a strong jammer there exists a coding scheme that will provide a channel capacity with an arbitrarily low error rate that is proportional to $\frac{S}{J_0}$, as long as we have sufficient bandwidth.

8.4.1 Direct Sequence Spread Spectrum

In chapter 2 we touched on DSSS systems and defined the processing gain of the system as $G_p = \frac{R_p}{R}$ where R_p is the chip rate and R is the underlying data rate. This definition should more correctly be $G_p = \frac{W_{ss}}{W_{min}}$ where W_{ss} is the spread bandwidth and W_{min} is the bandwidth of the data transmission, but the bit rate version is more convenient and assumes a spectral efficiency of 1 bps/Hz. A derivation of these formulae can be found in standard communication texts such as [2].

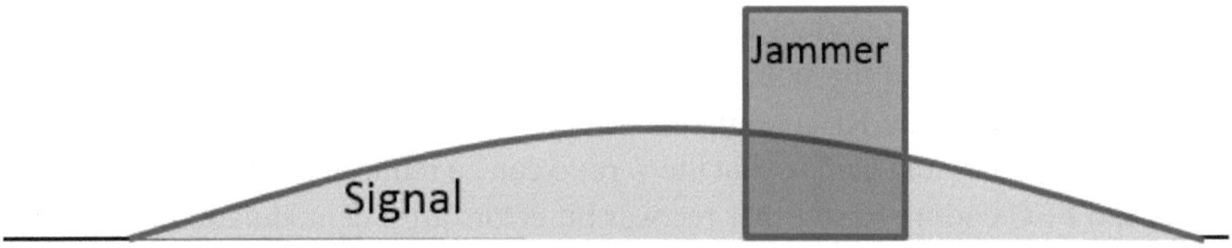

(a) Narrow band jamming spectrum

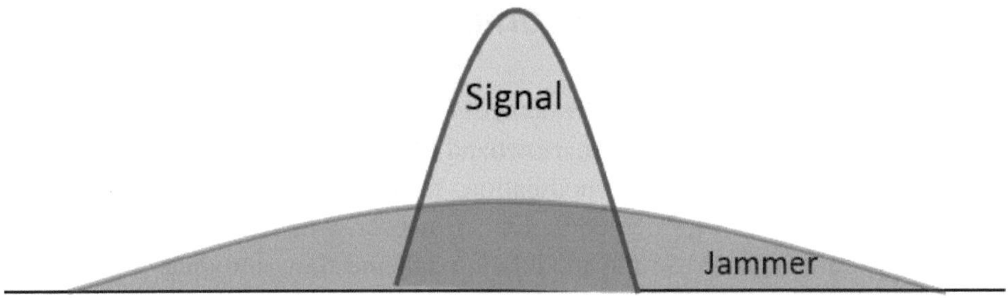

(b) Spectrum after de-spreading

Figure 8.4.1: DSSS spreading effect on narrow band jamming

From the jammer's perspective a simple narrow band jamming approach, such as the one described in section 8.1.2, will no longer be effective because, as shown in figure 8.4.1, only a relatively small portion of the signal will be corrupted. The processing gain is effectively reducing the jammer to signal power ratio. From the jammer's perspective a decision must be made on whether to focus the jammer power on a narrowband section of the waveform, or whether to target the entire bandwidth, W_{ss}, which spreads the jammer power much more thinly. In either case the jammer power spectral density, J_0, is the same.

Recognising that the jammer is now the dominant noise source we can, for digital systems, define the principal signal to noise ratio metric as $\frac{E_b}{J_0}$, where J_0 is the jammer power spectral density given by $J_0 = \frac{J}{W_{ss}}$. Hence,

$$\frac{E_b}{J_0} = \frac{S/R}{J/W_{ss}} = \frac{W_{ss}/R}{J/S} = \frac{G_p}{J/S}$$

and therefore,

$$\frac{J}{S} = \frac{G_p}{E_b/J_0} \quad (8.4.1)$$

$$= G_p dB - \left(\frac{E_b}{J_0}\right) dB \quad (8.4.2)$$

In the previous example $\frac{J}{S} = 13\text{dB}$ and therefore $\frac{E_b}{J_0} = G_p - 13\text{dB}$. For a digital voice application an acceptable probability of bit error may be $P_b = 10^{-4}$ which, from figure 8.3.1, requires an $\frac{E_b}{J_0} = 8.5$ dB. Therefore a processing gain of $G_p = 21.5$ dB is required, or a factor of $g_p = 141$. Tactical digital voice usually operates at R=16 kbps and therefore the spread bandwidth required is $W_{ss} = G_p \times R = 2.26$ MHz, or a chip rate, $R_p = 2.26$ Mbps. Note that the US MUOS UHF satellite system uses a DSSS approach with a chip rate of 3.84 Mbps.

8.4.2 Frequency Hopping

Frequency hopping systems are typically implemented using non-coherent FSK as the underlying modulation scheme. A slow hopping scheme transmits one or more symbols per hop, whereas a fast hopping scheme transmits one symbol over multiple hops. The processing gain for a frequency hopping system is still expressed as the ratio of the spread bandwidth to the information bandwidth, $G_p = \frac{W_{ss}}{W_{min}}$, but the calculation of W_{ss} is slightly different depending on whether a fast or slow hopping system is being used.

In DSSS systems the chip rate is the shortest transition in the system. A similar concept applies in frequency hopping systems, but the shortest transition changes between slow hopping systems and fast hopping systems. In a slow hopping system the chip rate is the symbol rate, but in a fast hopping system the chip rate is the hop rate, because it is faster than the symbol rate.

A t-bit pseudo-random generator is clocked at the hopping rate, R_c, to generate one of 2^t random binary numbers. A frequency synthesiser maps each number to one of the 2^t pre-defined hopping channels, as shown in figure 8.4.2. The receiver uses the same pseudo-random sequence in order to follow the hopping pattern.

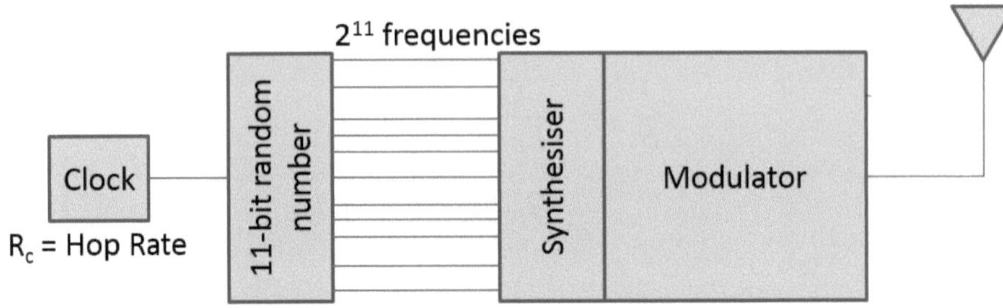

Figure 8.4.2: Frequency hopping block diagram

Slow Hopping

Consider a binary FSK slow hopping system with an information rate of R_b bits per second. The symbol rate is R_s symbols per second, with one bit transmitted per symbol, and one symbol is transmitted per hop. The bandwidth of the information signal is $W_{min} = R_s$Hz.

A t-bit frequency synthesiser can generate 2^t hopping frequencies, and each frequency channel must be separated by W_{min} Hz, as shown in figure 8.4.3, in order to satisfy the orthogonality requirements of FSK. Therefore, the spread bandwidth will be $W_{ss} = 2^t.W_{min}$ and the processing gain will be

$$G_p = \frac{W_{ss}}{W_{min}} = \frac{2^t W_{min}}{W_{min}} = 2^t.$$

That is, the processing gain in this slow hopping system is equal to the number of frequency channels.

Fast Hopping

In a fast hopping system the hop rate is faster than the symbol rate, which forces an increase in the frequency separation between the FSK tones in order to maintain the orthogonality constraint. Therefore the chip rate is considered to be the hop rate, rather than the symbol rate which was used for slow hopping systems.

Consider a fast hopping system that makes k hops per symbol. If W_{min} is the bandwidth required for the transmission of one data symbol, then the bandwidth required for the transmission of one hop (which has a duration of $\frac{T_s}{k}$) will be kW_{min}. As before, a t-bit synthesiser can generate 2^t hopping fre-

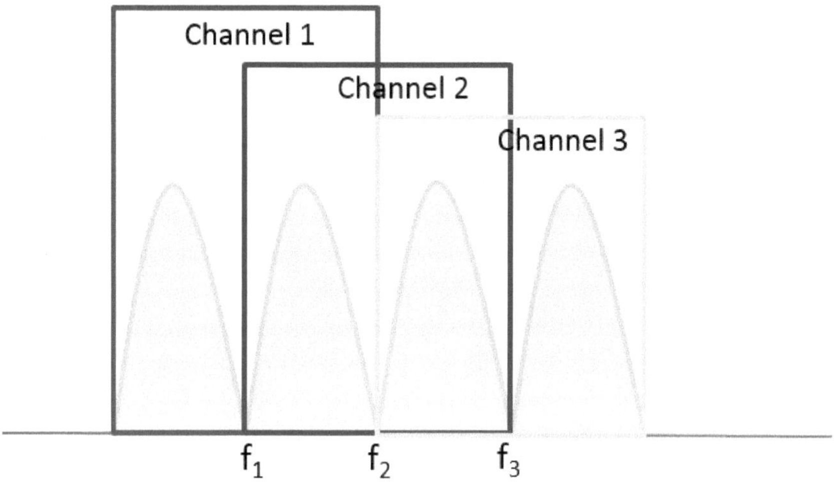

Figure 8.4.3: Channel separation of a slow hopping system

quencies, but each frequency channel must now be separated by kW_{min} Hz, as shown in figure 8.4.4, in order to satisfy the orthogonality requirements of FSK. Therefore, the spread bandwidth will be $W_{ss} = 2^t.kW_{min}$ and the processing gain will be

$$G_p = \frac{W_{ss}}{W_{min}} = \frac{2^t kW_{min}}{W_{min}} = k2^t.$$

That is, the processing gain in this fast hopping system is k times greater than the slow hopping version.

Example

The SINCGARS radio transmits 16 kbps in a 25 kHz channel using BFSK. In frequency hopping mode the radio hops at 111 hops per second over the 2,320 channels between 30 and 87.975 MHz. This is a slow hopping algorithm and therefore the processing gain is given by $G_p = 2320 = 33.7$ dB.

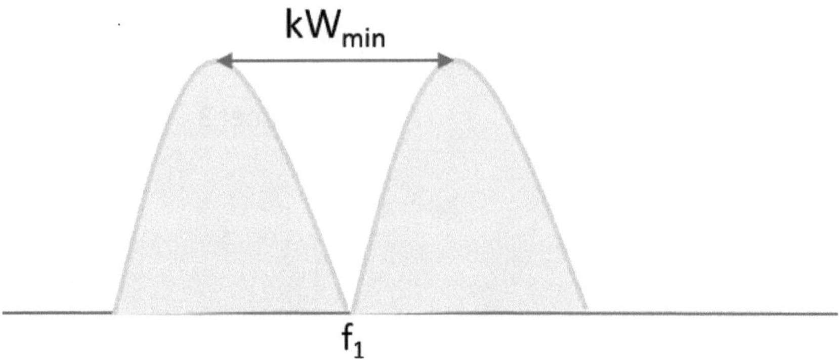

Figure 8.4.4: Channel separation of a fast hopping system

References

[1] David L. Adamy. *EW 103: Tactical battlefield communications electronic warfare*. Artech House, 2009.

[2] Bernard Sklar. *Digital Communications: Fundamentals and Applications*. Prentice-Hall, 1988.

Appendix 1 - Average Collection Area of a Receive Antenna

The Effective Area of a Receiving Antenna

The total spectral power collected by a receiving antenna, P_ν, is a product of its effective collecting area, A_e, and the power flux density, S, arriving at the receiver. Therefore, the effective area of the receiver can be expressed as:

$$A_e = \frac{P_\nu}{S_{(matched)}},$$

where $S_{(matched)}$ is the flux density in the *matched* polarisation. Polarisation is important because each antenna is designed to receive a signal with a particular polarisation; vertical or horizontal, or perhaps right or left circularly polarised signals. If the wave is essentially random, such as the noise generated by black body radiation, the two orthogonal components will vary rapidly in intensity but have equal powers when averaged over long times. Radiation with this characteristic is called unpolarised and for an unpolarised source,

$$S_{(matched)} = \frac{S}{2}.$$

The Law of Conservation of Energy requires that all lossless transmitting antennas have the same average power gain. Similarly, all lossless receiving antennas must have the same average collecting area and this average collecting area can be calculated by considering a thermodynamic experiment.

The antenna inside the left cavity is in thermodynamic equilibrium, at temperature T. It is connected through a transmission line to a matched resistor, R, in a second cavity at the same temperature. The connection contains a filter that passes only a narrow range of frequencies between and ν and $\nu + d\nu$. Since this entire system is in thermodynamic equilibrium, no net power can flow between the antenna and the resistor. Otherwise, one cavity would heat

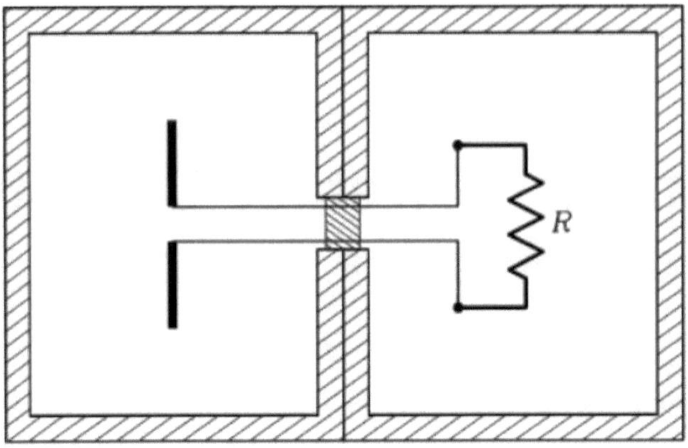

Figure 8.4.5: A cavity in thermodynamic equilibrium at temperature T containing a resistor R is coupled to an antenna, also at temperature T, through a filter passing frequencies in the range ν to $\nu + d\nu$.

up and the other would cool down, in violation of the second law of thermodynamics. Thus the total spectral power from all directions collected by the antenna is:

$$P_\nu = A_e S_{(matched)} = A_e \frac{S}{2} = \int_{4\pi} A_e(\theta, \phi) \frac{B_\nu}{2} d\Omega$$

which must equal the Nyquist spectral power $P_\nu = kT$ produced by the resistor. Using the Rayleigh-Jeans approximation $B_\nu = \frac{2kT}{\lambda^2}$ gives

$$P_\nu = \frac{2kT}{2\lambda^2} \int_{4\pi} A_e d\Omega = kT$$

and hence $\int_{4\pi} A_e d\Omega = \lambda^2$. The average collecting area is defined by

$$\langle A_e \rangle \equiv \frac{\int_{4\pi} A_e d\Omega}{\int_{4\pi} d\Omega}$$

and hence the effective collecting area of a receive antenna is independent of its radiation environment. Therefore, this result applies for any type of radiation, not just black body radiation, and we have the remarkable result true for all lossless antennas:

$$\langle A_e \rangle = \frac{\lambda^2}{4\pi}$$

This means that any antenna, from a short dipole to a 100m deep space tracking dish, has the same average collecting area A_e that depends only on wavelength. Importantly, in the case of an *isotropic* antenna, the effective collecting area in any direction equals the average collecting area:

$$A_e(\theta, \phi) = \langle A_e \rangle = \frac{\lambda^2}{4\pi}.$$

For other antenna types the gain of the antenna can be defined as the ratio of its effective area to the effective area of an isotropic radiator, $G = \frac{A_e}{\langle A_e \rangle}$. Hence, the effective area of an antenna is the product of the average collecting area for the particular frequency in use and the antenna gain in that direction,

$$A_e(\theta, \phi) = G(\theta, \phi)\frac{\lambda^2}{4\pi}.$$

Appendix 2 - Derivation of Power Limiting from Shannon's Formula

Shannon's formula indicates that channel capacity can not be continually increased by increasing bandwidth. In the limit, as $B \to \infty$, channel capacity becomes directly proportional to the signal to noise ratio.

$$\begin{aligned} H &= B \log_2 e \log_e \left(1 + \frac{C}{N}\right) \\ &= \frac{C}{N_0} \log_2 e \log_e \left(1 + \frac{C}{N}\right)^{\frac{N_0 B}{C}} \\ &= \lim_{B\to\infty} \frac{C}{N_0} \log_2 e \end{aligned}$$

Fact 1. *The McLauren expansion of* $\log_e (1 + x) = x - \frac{x^2}{2!} + \frac{x^3}{3!} - \ldots$

Proposition. $\lim_{B\to\infty} \log_e \left(1 + \frac{C}{N}\right)^{\frac{N_0 B}{C}} = 1$

Proof. Firstly,

$$\begin{aligned} \log_e \left(1 + \frac{C}{N}\right)^{\frac{N_0 B}{C}} &= \frac{N_0 B}{C} \log_e \left(1 + \frac{C}{N_0 B}\right) \\ &= \frac{N_0 B}{C} \left[\frac{C}{N_0 B} - \frac{1}{2}\left(\frac{C}{N_0 B}\right)^2 + \frac{1}{6}\left(\frac{C}{N_0 B}\right)^3 - \ldots\right] \\ &= 1 - \frac{C}{2 N_0 B} + \frac{C^2}{6 N_0^2 B^2} - \ldots \end{aligned}$$

Hence,

$$\begin{aligned}
\lim_{B\to\infty} \log_e \left(1+\frac{C}{N}\right)^{\frac{N_0 B}{C}} &= \lim_{B\to\infty} 1 - \frac{C}{2N_0B} + \frac{C^2}{6N_0^2B^2} - \ldots \\
&= 1
\end{aligned}$$

□

Appendix 3 - Radiation Hazard Analysis

ARPANSA Reference Levels

For practical purposes, radiation hazard is defined in the ARPANSA standard using a set of conservative Reference Levels that define safe exposure to electric and magnetic fields across the frequency range 3kHz to 300GHz. The measurable parameters that are used to define the levels are: electric field strength (**E** Volts/Metre), magnetic field strength (**H** Amps/Metre) or power density (P_d Watts/Square Metre) at the point of exposure. Where the point of exposure is in the *far field*, the simplest of these to calculate is power density which is related to electric field strength as:

$$|\mathbf{E}| = \sqrt{Z_0 P_d} \tag{8.4.3}$$

where

$Z_0 = \dfrac{|\mathbf{E}|}{|\mathbf{H}|}$ is the characteristic wave impedance ($120\pi \sim 377\ \Omega$ in free space)

Where the point of exposure is in the *near field* a simple power density calculation is no longer adequate. This is because the ratio of the magnitude of **E** and **H** is no longer constant. The solutions to Maxwell's equations show that for both **E** and **H** reactive fields exist alongside the more familiar radiative fields when the point of exposure is close to the antenna. The term "close" is measured with respect to the wavelength of the electromagnetic wave. Hence, the **E** field magnitude and the **H** field magnitude must be measured separately and compared with the Reference Levels in the standard.

Electromagnetic Waves

General

The general solution of Maxwell's equations for the electric and magnetic fields at a distance r from a small current element Idl is given by:

$$
\begin{aligned}
\mathbf{H} &= \frac{\mathrm{I}dl\sin\theta}{4\pi}\left(\frac{jk_0}{r}+\frac{1}{r^2}\right)e^{-jk_0 r}\mathbf{a}_\phi \\
\mathbf{E} &= \frac{\mathrm{jZ_0I}dl}{2\pi k_0}\cos\theta\left(\frac{jk_0}{r^2}+\frac{1}{r^3}\right)e^{-jk_0 r}\mathbf{a}_r \\
&\quad -\frac{\mathrm{jZ_0I}dl}{4\pi k_0}\sin\theta\left(-\frac{k_0^2}{r}+\frac{jk_0}{r^2}+\frac{1}{r^3}\right)e^{-jk_0 r}\mathbf{a}_\theta
\end{aligned}
$$

where

$$
\begin{aligned}
k_0 &= \frac{2\pi}{\lambda} \\
j &= \sqrt{-1}
\end{aligned}
$$

and the vector directions r, θ, ϕ define the location of the point of exposure in space. For simplicity we will assume that $\theta = \frac{\pi}{2}$, which means that the direction to the point of exposure is orthogonal to the orientation of the current element dl. Thus the equations simplify to:

$$
\begin{aligned}
\mathbf{H} &= \frac{\mathrm{I}dl}{4\pi}\left(\frac{jk_0}{r}+\frac{1}{r^2}\right)e^{-jk_0 r}\mathbf{a}_\phi \\
\mathbf{E} &= -\frac{\mathrm{jZ_0I}dl}{4\pi k_0}\left(-\frac{k_0^2}{r}+\frac{jk_0}{r^2}+\frac{1}{r^3}\right)e^{-jk_0 r}\mathbf{a}_\theta
\end{aligned}
\qquad (8.4.4)
$$

The Far Field

Clearly the **E** and **H** fields have both radiative (ie, real) components and reactive (ie, imaginary) components. If $r \gg \lambda$ then the point of exposure is in the far field and the linear terms $\left(\frac{1}{r}\right)$ dominate, which produces the predictable electromagnetic wave parameters. In the far field the electric and magnetic fields can be approximated as:

$$\begin{aligned}
\mathbf{E} &= \mathrm{jZ_0I}dlk_0\frac{e^{-jk_0r}}{4\pi r}\mathbf{a}_\theta \\
\mathbf{H} &= j\mathrm{I}dlk_0\frac{\mathrm{e^{-jk_0r}}}{4\pi r}\mathbf{a}_\phi
\end{aligned}$$

The power density of the wave is given by the Poynting vector $\mathbf{S} = \frac{1}{2}\mathbf{E}\times\mathbf{H}$ and is real, which reflects the radiative nature of the wave in the far field. The magnitude of the **E** field is given by:

$$|\mathbf{E}| = \frac{\mathrm{Z_0I}dlk_0}{4\pi r}$$

However, the magnitude of the **E** field is also given in terms of the transmitted power by (8.4.3). Hence,

$$\begin{aligned}
\frac{\mathrm{Z_0I}dlk_0}{4\pi r} &= \sqrt{Z_0P_d} \\
&= \sqrt{\frac{Z_0P_t}{4\pi r^2}} \\
&= \frac{\sqrt{4Z_0P_t}}{4\pi r}
\end{aligned}$$

and

$$\mathrm{Z_0I}dlk_0 = \sqrt{4Z_0P_t} \tag{8.4.5}$$

where P_t is the antenna transmitted power.

The Near Field

In the near field the $\frac{1}{r}$ components are still present, but for small r they are dominated by the $\frac{1}{r^3}$ and $\frac{1}{r^2}$ components. Note that the power density $\mathbf{S} = \frac{1}{2}\mathbf{E}\times\mathbf{H}$ is purely imaginary for some component combinations, which represents reactive power in the near field. Even in the near radiative field these additional components can contribute significant energy.

Using equation (8.4.4), and ignoring phase relationships, the magnitude of the electric field in the near field is roughly:

$$
\begin{aligned}
|\mathbf{E}| &= \frac{\mathrm{Z_0 I} dl k_0}{4\pi k_0}\left(\frac{k_0}{r}+\frac{1}{r^2}+\frac{1}{k_0 r^3}\right)\\
&= \frac{\sqrt{4Z_0 P_t}}{4\pi k_0}\left(\frac{k_0}{r}+\frac{1}{r^2}+\frac{1}{k_0 r^3}\right)\\
&= \frac{\sqrt{4Z_0 P_t}}{4\pi}\left(\frac{1}{r}+\frac{1}{k_0 r^2}+\frac{1}{k_0^2 r^3}\right)\\
&= \frac{\sqrt{4Z_0 P_t}}{4\pi}\left(\frac{1}{r}+\frac{\lambda}{2\pi r^2}+\frac{\lambda^2}{4\pi^2 r^3}\right)
\end{aligned}
$$

Example

Using an example frequency of $f = 10\text{MHz}$ then the wavelength is $\lambda = 30\text{m}$ and the near field boundary occurs at $\frac{\lambda}{2\pi} = 4.7\text{m}$. If a HF antenna is located at a distance to the cabin of 2.5m, it places the cabin well inside the near field boundary. The approximate magnitude of the electric field at this distance is

$$
\begin{aligned}
|\mathbf{E}| &= \frac{\sqrt{4\times 377\times 150}}{4\pi}\left(\frac{1}{2.5}+\frac{30}{2\pi(2.5)^2}+\frac{30^2}{4\pi^2(2.5)^3}\right)\\
&= 99.27\text{Vm}^{-1}
\end{aligned}
$$

Some of this energy is not radiative. Instead, the reactive field exchanges the energy with the antenna through each cycle, with only the $\frac{1}{r}$ radiative elements transmitting energy into the far field. However, humans have a measure of capacitance (a reactive parameter) such that if a person is standing near the antenna then some of the energy in the reactive field will be exchanged with the capacitive nature of the human body. This exposes the person to electromagnetic radiation in excess of the ARPANSA recommended limit, which is only 61.4 Vm^{-1} at 10MHz.

Figure 8.4.6 illustrates the effect of the $\frac{1}{r^2}$ and $\frac{1}{r^3}$ components on electric field strength magnitude as the distance changes from the antenna. To aid visibility the log of the field strength has been plotted on the y-axis. The upper (blue) line indicates the electric field strength when the contribution from all components is included. The lower (red) line indicates the electric field strength when only the $\frac{1}{r}$ component is included. It can be seen that very close to the antenna the higher order components produce extremely high electric field

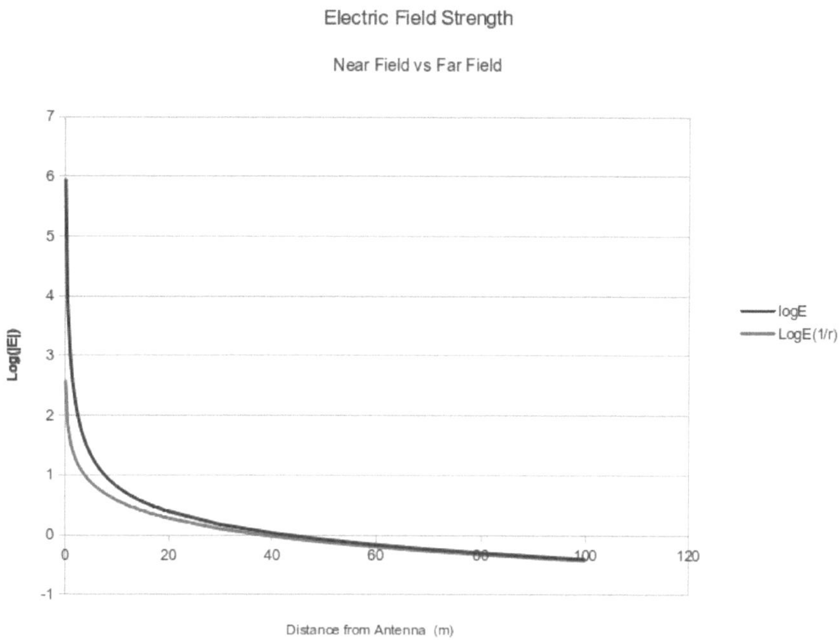

Figure 8.4.6: Electric Field Strength - Near vs Far Field

strength values; at 0.1 metres the value is around 800,000 V/m. At a distance of around one wavelength (30 metres) the effect of the higher order components disappears and the total electric field strength is essentially the same as the contribution from the $\frac{1}{r}$ component. This is the far field response.

Appendix 4 - Orthonormality of PSK Basis Functions

In section 2.2.2 the basis functions for PSK modulation are given as

$$\begin{aligned}\phi_1(t) &= \left(\frac{2}{T_s}\right)^{\frac{1}{2}} \cos(\omega_c t) \\ \phi_2(t) &= \left(\frac{2}{T_s}\right)^{\frac{1}{2}} \sin(\omega_c t),\end{aligned}$$

where T_s is the symbol period and $\omega_c = 2\pi f_c$ is the carrier frequency. The claim is made that these functions are orthonormal for the inner product defined as

$$\int_0^{T_s} \phi_i \phi_j dt = \delta_{ij} = \begin{cases} 1, & i = j \\ 0, & i \neq j \end{cases}.$$

However, this claim is only strictly true if the carrier frequency, f_c, is an integer multiple of the data rate, $R_b = \frac{1}{T_s}$. For example,

$$\begin{aligned}\int_0^{T_s} \phi_2 \phi_2 dt &= \int_0^{T_s} \frac{2}{T_s} \sin^2(\omega_c t) \\ &= \frac{2}{T_s} \int_0^{T_s} \frac{1 - \cos(2\omega t)}{2} dt \\ &= \frac{2}{T_s} \int_0^{T_s} \frac{1}{2} dt - \frac{1}{T_s} \int_0^{T_s} \cos(2\omega t) dt \\ &= \frac{2}{T_s} \left[\frac{T_s}{2}\right] - \frac{1}{2\omega T_s} \sin(2\omega T_s) \\ &= 1 - \frac{\sin(2\omega T_s)}{2\omega T_s}. \qquad (8.4.6)\end{aligned}$$

If $f_c = mR_b$, where m is an integer, then $2\omega T_s = \frac{4\pi T_s}{t_c} = 4\pi m$ and the second term in equation 8.4.6 is zero. However, if $f_c \neq mR_b$ then the second term will not be zero, but will be of the form $\frac{\sin(a)}{a}$. Since $\lim_{a\to\infty} \frac{\sin(a)}{a}. = 0$, this term will be negligible as long as $\frac{T_s}{t_c}$ is large, or in other words, as long as $f_c \gg R_b$, which is usually the case.

Bibliography

[1] 4RF. Aprisa xe application note: Military los tactical radio relay, July 2012. Issue 2.0.0.

[2] D. Adamy. *EW 102 A second course in electronic warfare.* Artech House, 2003.

[3] D. L. Adamy. *EW 103: Tactical battlefield communications electronic warfare.* Artech House, 2009.

[4] ADF. *Australian Defence Doctrine Publication 3.13 Information Operations.* Defence Publishing Service Canberra 2600, 2006.

[5] D. of the Army. Tb 11-5825-298-10-3 enhanced position location reporting system (eplrs) planner's manual. Tech. rep., US Army, 2009.

[6] U. Army. *FM 6-02.53 Tactical Radio Operations.* Department of Defense, 2009.

[7] D. Axe. Failure to communicate: Inside the army's doomed quest for the 'perfect' radio, January 2012, http://www.publicintegrity.org/2012/01/10/7816/failure-communicate-inside-armys-doomed-quest-perfect-

[8] K. Bergmann. Battle management systems comes together. *Asia Pacific Defence Reporter* October:12–13, 2012, http://www.asiapacificdefencereporter.com/articles/261/Battle-management-systems-comes-together.

[9] C. S. Bey. Airborne tactical data network gateways: Evaluating eplrs' ability to integrate with wireless meshed networks. Master's thesis, Naval Postgraduate School, Monterey, California, 2005.

[10] R. Bowman. Improved performance for iintegrate voice and data in a tactical packet network. *Military Communications*, 1998.

[11] R. Bowman and H. Russell. Route selection for data packets in a tactical packet radio network using integrated voice and data transmissions. *Military Communications*, 1999.

[12] R. Bowman and H. Russell. Channel access for simultaneous voice and data transmission in a multiple-hop packet radio network. *Military Communications*, 1998.

[13] J. T. N. Center. Our products, Accessed 17 Mar 2014, `http://jtnc.mil/Pages/Products.aspx`.

[14] R. Connor. Vetronics standards and guidelines - issue 3, July 2009, `http://www.vsi.org.uk/Standards_-_Metrics/SG_2007_Issue3.pdf`.

[15] H. Corporation. *Radio communications in the digital age*, vol. 1. Harris Corporation, 2 edition, 2005.

[16] N. G. Corporation. Understanding voice and data link networking. Tech. rep., Northrop Grummon Corporation, Dec 2013.

[17] R. Corporation. Raytheon technical workbook eplrs system technical description. Tech. rep., Raytheon Corporation, 2008.

[18] G. Ebbutt. Digitisation of the battlefield. *European Security and Defence Press Association*, 2013, `http://www.esdpa.org/2013/10/digitisation-of-the-battlefield/`.

[19] G. F. Elmasry. The progress of tactical radio from legacy systems to cognitive radios. *IEEE Communications Magazine* 51:50–56, October 2013.

[20] G. F. Elmasry. *Tactical Wireless Communications and Networks: Design Concepts and Challenges*. Wiley, 2012.

[21] G. F. Elmasry. A comparative review of commercial vs tactical wireless networks. *IEEE Communications Magazine* 48:54–59, 2010.

[22] J. W. Fry. An analysis of mobile ad-hoc network performance to recommend a basis of issue for the u.s. army nett warrior system. Master's thesis, Naval Postgraduate School, Monterey, California., September 2010.

[23] L. H. G. Petch, P Straughair. Occasional paper: The impact of tactical communications on future military effectiveness. Tech. rep., Raytheon Australia Pty Ltd, 2012.

[24] GlobalSecurity.org. Joint tactical radio system: Programmable, modular communications, Accessed 14 March 2014, `http://www.globalsecurity.org/military/systems/ground/jtrs.htm`.

[25] B. J. Hamilton. Sincgars system improvement program (sip) specific radio improvements. *Tactical Communications Conference*, pp. 397–406, 1996. 10.1109/TCC.1996.561109.

[26] K. Kang. Tactical internet overview. *Korea Internet Conference*, June 2008.

[27] R. Larson and B. Edwards. *Elementary Linear Algebra*. D.C. Heath and Company, 1988.

[28] T. Mann. Jtrs/win-t: networking waveform quick reference sheets. *Army Communicator* 33(3):72–76, June 2008.

[29] C. Matassa. Comparing the capabilities and performance of the ultra high freqency follow-on system with the mobile user objective system. Master's thesis, Naval Postgraduate School, Monterey, California, June 2011.

[30] H. Nyquist. Certain topics of telegraph transmission theory. *Transactions of the American Institute of Electrical Engineering* 47:617–644, April 1928.

[31] L. W. R. Schulman, R. Snyder. Sincgars internet controller - heart of the digitized battlefield. *Proceedings of the 1996 Tactical Communications Conference*, pp. 417–421, April 1996.

[32] Raytheon. Raytheon demos radio-based network in afghanistan, June 2013, `http://www.defencetalk.com/raytheon-demos-radio-based-network-in-afghanistan-4811`

[33] M. Ryan and M. Frater. *Battlefield Communications Systems*. Argos Press, 2007.

[34] P. Sass. Communications networks for the force xxi digitized battlefield. *Mobile Networks and Applications* 4:139–155, 1999.

[35] C. E. Shannon. A mathematical theory of communication. *Bell System Technical Journal* 27:379–423, 623–656, 1948.

[36] B. Sklar. Reed-solomon codes, 2002, `http://ptgmedia.pearsoncmg.com/images/art_sklar7_reed-solomon/elementLinks/art_sklar7_reed-solomon.pdf`.

[37] B. Sklar. *Digital Communications: Fundamentals and Applications*. Prentice-Hall, 1988.

[38] D. Tharp and L. Wallace. Enhanced position location reporting system: Legacy system provides new technology for warfighters. *Space*

and Naval Warfare Systems Center San Diego Biennial Review 1:206–211, 2003.

[39] A. Toffler. *Future Shock.* Random House, 1970.

[40] A. Viterbi. Spread spectrum communications - myths and realities. *IEEE Communications Magazine* 17(3):11–18, May 1979.

[41] A. Viterbi. Error bounds for convolutional codes and an asymtotically optimum decoding algorithm. *IEEE Transactions on Information Theory* IT13:260–269, April 1967.

[42] H. S. W. Pritchard and R. Nelson. *Satellite Communications Systems Engineering.* Prentice Hall, 1993.

[43] A. Whalen. *Detection of Signals in Noise.* Academic Press, Inc, 1971.

[44] T. Withington. Middle managers: Battle management systems. *Asian Military Review*, 2014, `http://www.asianmilitaryreview.com/middle-managers/`.

[45] Def stan 00-82: Vetronics infrastructure for video over ethernet (viveo) issue 2.

Index

Printed by Libri Plureos GmbH in Hamburg,
Germany

9 780992 529901